AIChE Equipment Testing Procedure

Mixing Equipment (Impeller Type)

3rd Edition

Prepared by the Equipment Testing Procedures Committee
of the American Institute of Chemical Engineers

AIChE

American Institute of Chemical Engineers
www.aiche.org

3 Park Avenue, New York, NY 10016-5991
www.aiche.org
ISBN 0-8169-0836-2
Pub. E-30

Cover Design: Beth Shery Sisk

Printed in the United States of America.

It is sincerely hoped that the information presented in this document will lead to even more impressive performance by the chemical processing and related industries. However, the American Institute of Chemical Engineers, its employees and consultants, its officers and directors, Equipment Testing Procedures Committee members, their employers, and their employers' officers and directors disclaim making or giving any warranties or representations, express or implied, including with respect to fitness, intended purpose, use or merchantability and/or correctness or accuracy of the content of the information presented in this document. Company affiliations are shown for information only and do not imply approval of the procedure by the companies listed. As between (1) the American Institute of Chemical Engineers, its employees and consultants, its officers and directors, Equipment Testing Procedures Committee members, their employers, and their employers' officers and directors, and (2) the user of this document, the user accepts any legal liability or responsibility whatsoever for the consequence of its use or misuse.

Approved by AIChE's Chemical Engineering Technology Operating Council on June 3, 2000.

Members Participating in Second Edition

C. N. Carpenter
R. A. Coble
C.K. Coyle
D.S. Dickey
J.B. Fasano
C.M. Garrison
J.B. Gray
W.D. Ramsey
R.N. Salzman
W.D.R. Short
J.A. Von Essen

Second Edition officially approved by AIChE Council on November 24, 1987 for publication

Members Participating in First Edition

R.L. Bates
W.G. Canham, Jr.
J.R. Connolly
J.B. Gray
R.H. Jebens
S.L. Lopata
E.J. Lyons
C.M. Nelson
H.F. Nolting
J.Y. Oldshue
R.B. Olney
T. Vermeulen
S.S. Weidenbaum

First Edition officially approved by AIChE Council in 1959 for publication.
Edition revised, January 1965

Contents

Figures

100.0 Purpose and Scope

101.0 *Purpose*

This procedure offers methods of conducting and interpreting performance tests on impeller-type mixing equipment.

These tests may be conducted to determine process performance, mechanical reliability, or suitability of equipment for the intended use. Since the correct identification of the "real" problem can be the most difficult part of the tests conducted on mixing equipment, several procedures follow troubleshooting tactics.

Tests may be conducted to determine scale-up or scale-down criteria and to collect data for equipment sizing and design.

Many reasons exist for conducting performance tests. The methods presented here should be generally applicable to most situations. Care should be taken to set testing priorities and to select the most suitable methods for a given situation.

102.0 *Scope*

Rather than specific instructions, a collection of techniques is presented to guide the user. Emphasis is placed on practical methods that are likely to produce reliable results.

This procedure includes widely accepted nomenclature and definitions to assist in the collection of data and communication of results. General methods are provided for collecting and analyzing process results, but because of the enormous variety of possible applications for impeller-type mixing equipment, little specific detail is included.

Many useful indirect measures of process conditions involve mechanically related observations. Because mechanically sound equipment is necessary for successful process operation, many aspects of the testing are mechanical. Observations of mechanical operation are also essential for long equipment life and personnel safety.

103.0 *Liability*

It is sincerely hoped that the information presented in this document will lead to even more impressive performance by the chemical processing and related industries. However, the American Institute of Chemical Engineers, its employees and consultants, its officers and directors, Equipment Testing Procedures Committee members, their employers, and their employers' officers and directors disclaim making or giving any warranties or representations, express or implied, including with respect to fitness, intended purpose, use or merchantability and/or correctness or accuracy of the content of the information presented in this document. Company affiliations are shown for information only and do not imply approval of the procedure by the companies listed. As between (1) the American Institute of Chemical Engineers, its employees and consultants, its officers and directors, Equipment Testing Procedures Committee members, their employers, and their employers' officers and directors, and (2) the user of this document, the user accepts any legal liability or responsibility whatsoever for the consequence of its use or misuse.

200.0 Definition and Description of Terms

201.0 *Introduction*

Impeller-type mixing equipment includes a variety of rotating equipment used for fluid processing. No single description can provide complete information about all types of equipment. Usually, impeller-type mixing equipment includes both the rotating mixing equipment and the tank in which it is used. The fluid in the tank is also an important consideration in any testing procedure.

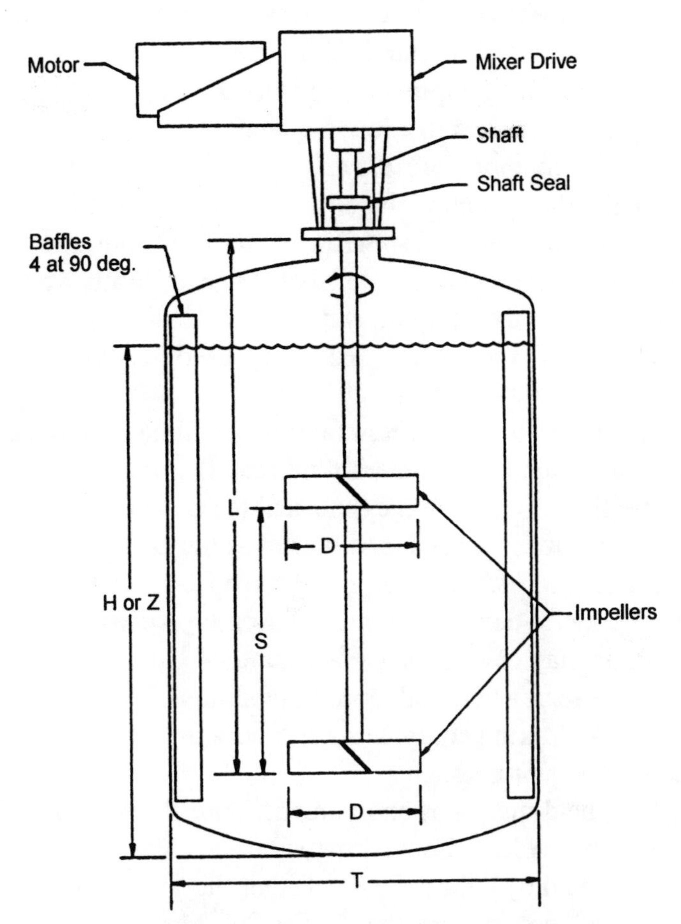

Fig. 201.1 Impeller Mixing Equipment

The terms, mixing and agitation, are usually used almost interchangeably. Both terms generally describe the results of liquid motion produced by a rotating impeller. However, mixing may be used to refer more specifically to the blending of dissimilar liquids or components. Agitation may be used more generally to describe the motion of fluids for blending, solids suspension, gas dispersion, etc. In this procedure, the terms mixing and agitation will be used to interchangeable to describe equipment and processes.

202.0 *Mixing Equipment*

An impeller-type mixer or agitator can be defined as equipment for blending and mixing of liquids with liquids, liquids with dispersed solids, or liquids with dispersed gases, or liquids with both dispersed solids and gases, provided a liquid phase is continuous. A rotating impeller provides both a thrusting and shearing action to the fluid in a vessel, which results in both flow and turbulence. The equipment takes many forms, but common to each is a device (impeller) attached to a rotating shaft.

202.1 *Equipment Configurations*

Usually, the system includes the impeller-type mixer, the vessel, all internal accessories, and sometimes auxiliary equipment. The impeller mixer usually consists of five (5) basic components: a motor (prime mover), a mixer (agitator) drive that reduces speed and increases torque (not always required), a shaft seal (used only with closed tanks), a shaft and impeller(s). See Fig. 201.1

202.1.1 The most common impeller mixer arrangement is the center-mounted, top-entering mixer. The shaft is vertical at the centerline of an upright cylindrical tank. Various types of impellers may be used, and baffles at the tank walls are usually necessary to prevent swirling of the contents and excessive vortex formation. Such equipment is extremely versatile, and the tank volume may be less than a few hundred gallons (one cubic meter) to more than a hundred thousand gallons (five hundred cubic meters). Mixing equipment for laboratory applications can be used in smaller volumes.

The dimensions and nomenclature shown in Fig. 201.1 are basic and typical. Some dimensions, such as liquid level may be represented by either H or Z. Either the shaft length, L (Fig. 201.1) or the off-bottom location, C (Fig 202.1) may be used to identify the location of the lower impeller. The dimension C, shown in Fig. 202.1, is the off-bottom location of the lower impeller, not the clearance. Impeller locations are often referenced to the bottom center of the tank, since mounting location does not affect fluid motion.

Some manufacturers and researchers measure impeller location to the centerline of the impeller, while others measure to the bottom of the impeller. The distance from the bottom of the impeller to the tank bottom is the clearance. Yet because many tank bottoms are dished the minimum clearance may be near the impeller blade tip, not the center of the tank. For test purposes, the impeller location should be noted with clearly defined dimensions. If initial installation is important, impeller clearance needs to be adequate.

Additional dimensions may be needed to describe the tank completely. Dimensions for internals (coils, dip pipes, etc.), a heat transfer jacket, and nozzles on the top, side, or bottom of the tank should be recorded. For process description the tank diameter, T, is the inside diameter, although tank dimensions could use outside diameter, in which case wall thickness is important. Be sure that all dimensions are clearly described and documented.

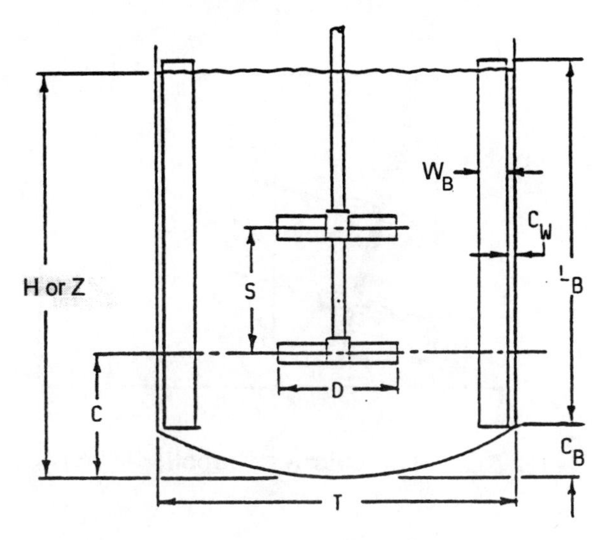

Fig. 202.1 Center Mount Impeller Mixer

202.1.2 Top-entering, angle-mounted mixers, eliminate the need for baffles by using the axial flow of the impeller(s) to counteract swirl. However, the weight of the shaft and impeller on an angle-mounted mixer applies a bending load to the drive, shaft, and support. Thus, most angle-mounted mixers are less than 5 hp (4 kW) and operate at speeds greater than 250 rpm (4 rps).

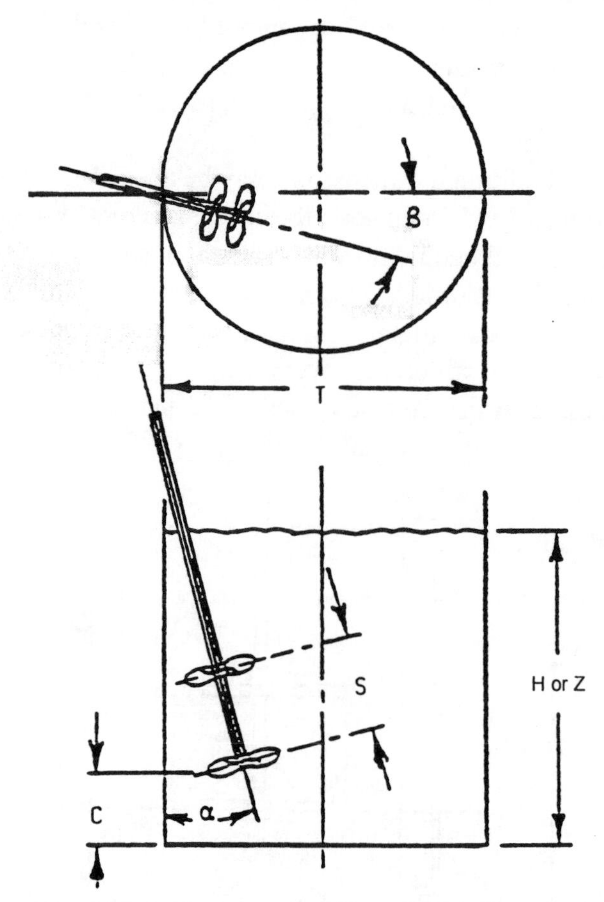

Fig. 202.2 Top-entering Propeller Mixer

202.1.3 Top-Entering, off-center mounted mixers may also reduce the swirl sufficiently to eliminate the need for baffles. The elimination of baffles may avoid product hang-up and make cleaning easier. However, off-center mounting creates larger hydraulic loads that may require a stronger shaft and drive, stiffer mounting, and/or a bottom steady bearing. The offset, *O*, from the centerline of the tank is typically 1/6 to 1/4 of the tank diameter, *T*.

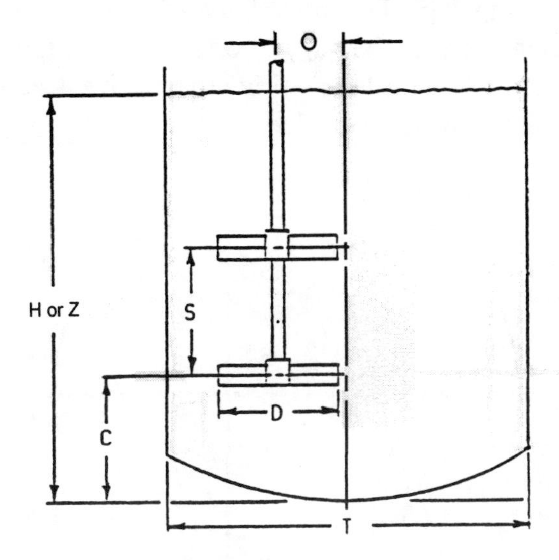

Fig. 202.3 Top-entering Off-Center Mount Mixer

202.1.4 Side-entering mixers are most often used in large, field-erected, storage tanks or concrete chests. The mixer is usually mounted near the bottom of the tank, where the off-bottom location, *C*, is about equal to the impeller diameter, *D*. Mounting the mixer near the bottom provides for the maximum variation in the liquid level. If the liquid level gets too close to the impeller, splashing will occur, which may cause severe loads on the mixer shaft and impeller. The mounting angles for a side-entering mixer are chosen to create the most effective flow pattern for the application. Some side-entering mixers have a swivel feature to redirect the flow when needed.

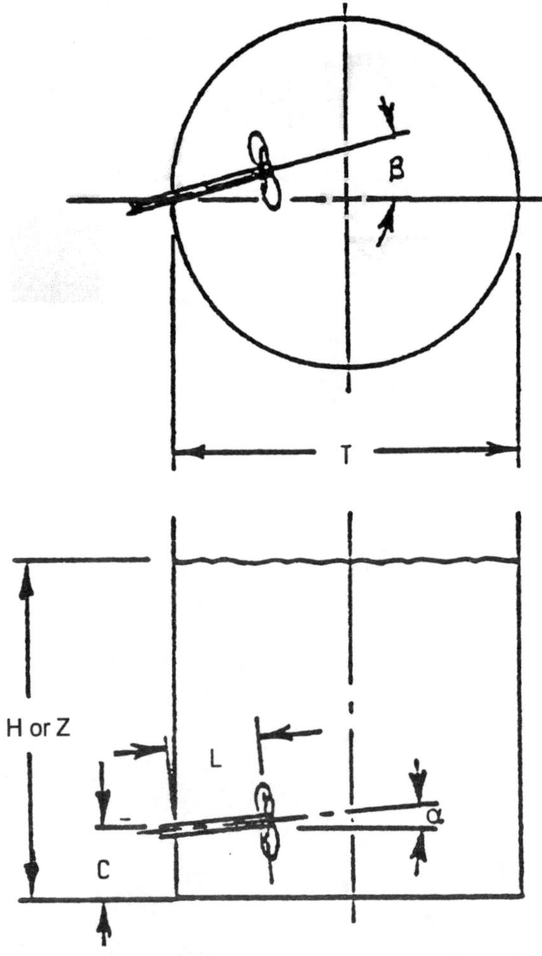

Fig. 202.4 Side-entering Mixer

202.1.5 Close-clearance impeller systems are a special case of the center-mounted mixers. The center-mounted mixers described in Sec. 202.1.1 typically have impeller diameters that are between 15 percent and 75 percent of the tank diameter. Close-clearance impellers are likely to be between 85 percent and 95 percent of the tank diameter. Some close-clearance impellers even have flexible scrapers that contact the tank wall as they rotate. Close-clearance impellers are typically used with viscous or non-Newtonian fluids.

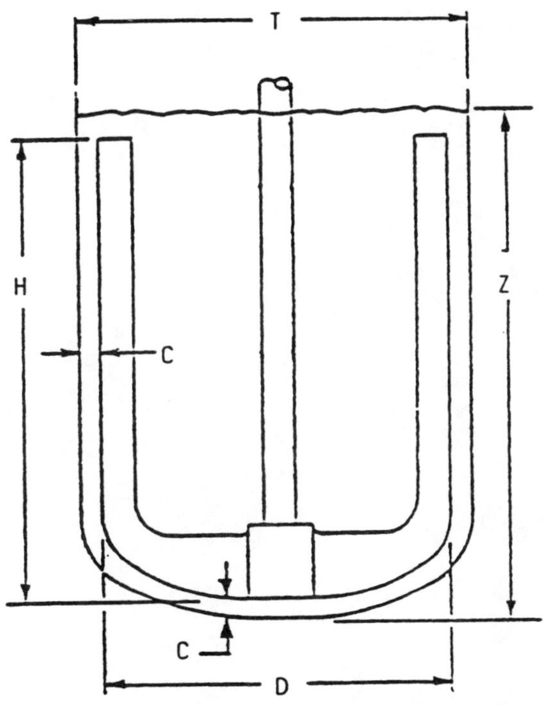

Fig. 202.5 Close Clearance Impeller Mixer

202.1.6 Additional mixer types include bottom-entering mixers and mixers mounted in draft tubes. Some tanks have multiple top or side-entering mixers and may have different impeller types on single or multiple shafts. Most of the test methods in this procedure can be modified to handle these other types of mixers.

202.1.7 For additional testing procedures for mixing equipment used with dry solids, paste, and dough, see Ref. 808.11

202.2 *Impellers*

An impeller is defined by a set of physical and geometric factors including diameter, number of blades, contour of blades (blade shape), width of blades, angle of blades, and thickness of blades.

Typical impeller types include radial-flow impellers, tangential-flow impellers, axial-flow impellers, anchors, augers (screws), and helixes. The term "turbine" is frequently used when referring to impellers with flat-plate-style blades. Examples of impellers are shown in Figs. 202.6 through 202.15.

202.2.1 **Straight-blade Turbine** - A straight blade turbine is a basic radial flow impeller. As the impeller rotates, the primary direction of discharge flow is radially outward from the axis of the shaft. As liquid circulates in the tank, flow returns to both the top and bottom of the impeller. Straight-blade turbines typically have two, four, or six blades, but any practical number of blades may be used. Typical blade widths are about 1/6 or 1/5 of the impeller diameter, but a wide range of widths is possible. Although usually flat and straight, the blades may be swept backward from the direction of rotation, making a **Curved-blade Turbine**. The curved blades may offer a different radial flow and may reduce the accumulation of fibrous or sticky material on the front of the blades.

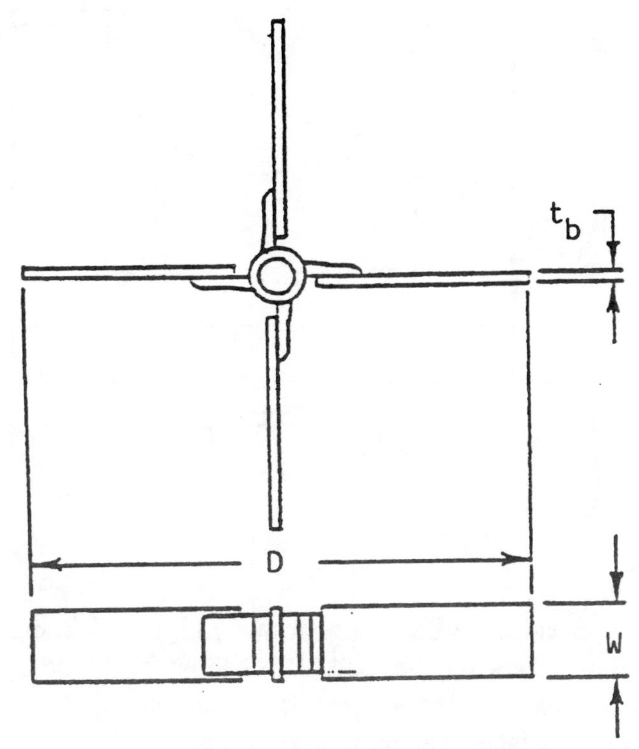

Fig. 202.6 Straight-blade Turbine

202.2.2 **Disc-style Turbine** - A disc-style turbine is another type of radial flow impeller, which creates flow patterns similar to the straight-blade turbine. The disc-style turbine has the blades mounted to a horizontal disc, instead or directly to the impeller hub. With bolted blades mounted on a disc, extra holes can make impeller diameter and/or number of blades adjustable. The disc may also provide a better radial flow pattern for gas dispersion.

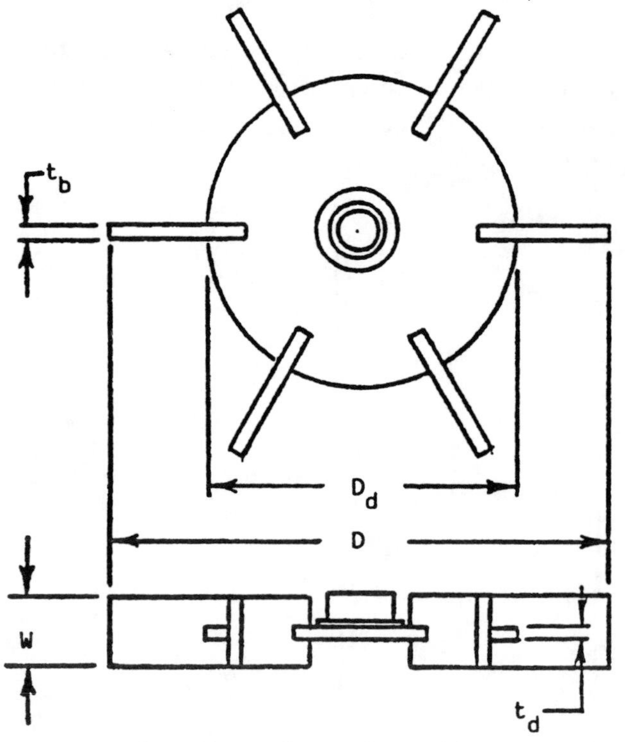

Fig. 202.7 Disc-style Turbine

Besides the flat blades shown in Fig. 202.7, blades with other cross sections may be used. Blades may be pitched. Blades may have a semicircular or parabolic cross section. The concave side of the blades faces the direction of rotation.

Disc-style impellers with blades having curved cross-sections can disperse more gas than impellers with similar projected-width flat blades. Blades with curved cross sections also experience less power reduction than straight blades for the same gas flow rates. The combination of these characteristics may provide better gas-liquid mass transfer with curved blades.

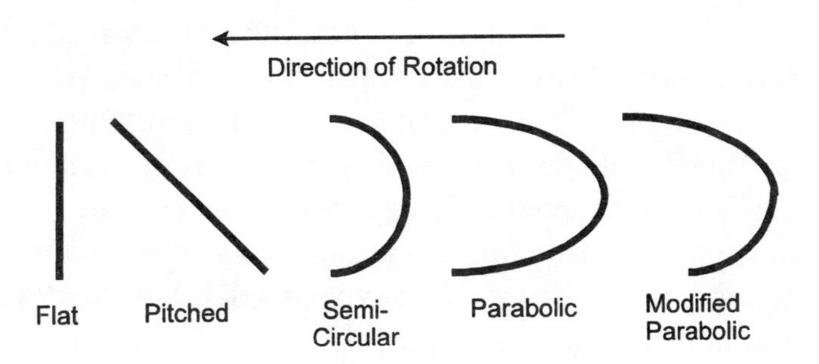

Fig. 202.8 Blade Shapes for Disc-style Turbines

202.2.3 **Pitched-blade Turbine** - A pitched-blade turbine is a simple variation from the straight-blade turbine, where the blades are mounted at an angle. The angle is normally measured from the horizontal, with 45 deg. angle blades the most common. Impellers with blades mounted at 30 deg. or even 60 deg. are also used. Most pitched-blade turbines are rotated so that the direction of pumping is downward. This type of axial flow creates good top-to-bottom motion in a tank, which results in good mixing. Simply inverting the impeller does not change the direction of the pumping action, only reversing the direction of rotation or changing the angle of the blades will make a pitched-blade up-pumping.

Impeller diameter can be measured either from tip to tip, D, or on the swept diameter of the blade circle, D_s. The straight diameter is adequate for description and fabrication of simple plate blades. The swept diameter, D_s, can be used to describe many blade shapes. The swept diameter includes the effects of blade shape, width and thickness. Swept diameter is an important dimension for installation, to assure fit through openings or clearance to tank internals.

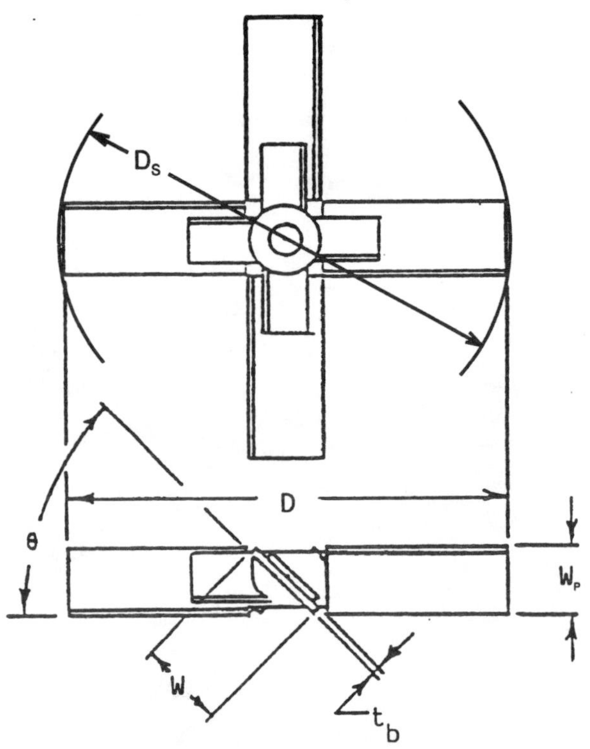

Fig. 202.9 Pitched-blade Turbine

When calculating power from a power number, one must know whether the tip diameter, D, or the swept diameter, D_s, was used in the correlation. Small diameter differences make large power differences (power is proportional to the diameter to the fifth exponent).

202.2.4 **Axial-flow Impeller** - A class of axial flow impellers, also called hydrofoil impellers, with lower power numbers than pitched-blade turbines, have been designed to produce high flow with low turbulence. These impellers often provide efficient liquid blending and solids suspension. Most of these axial flow impellers are proprietary, with each manufacture offering a unique design. A few of the designs or design characteristics may be covered by patents. Most of these impellers have three or four blades, made of rolled or bent, metal plate. The blades are bolted or welded to a central hub that mounts on the shaft. The width and shape of the blades may vary considerably depending on the manufacturer and application.

The impeller design shown in Fig. 202.10 is typical of the high flow, low power number designs. Impellers with four blades and/or wider blades, or with steeper blade angles, are usually less efficient in producing flow, but offer other process advantages. No one design is best for all applications.

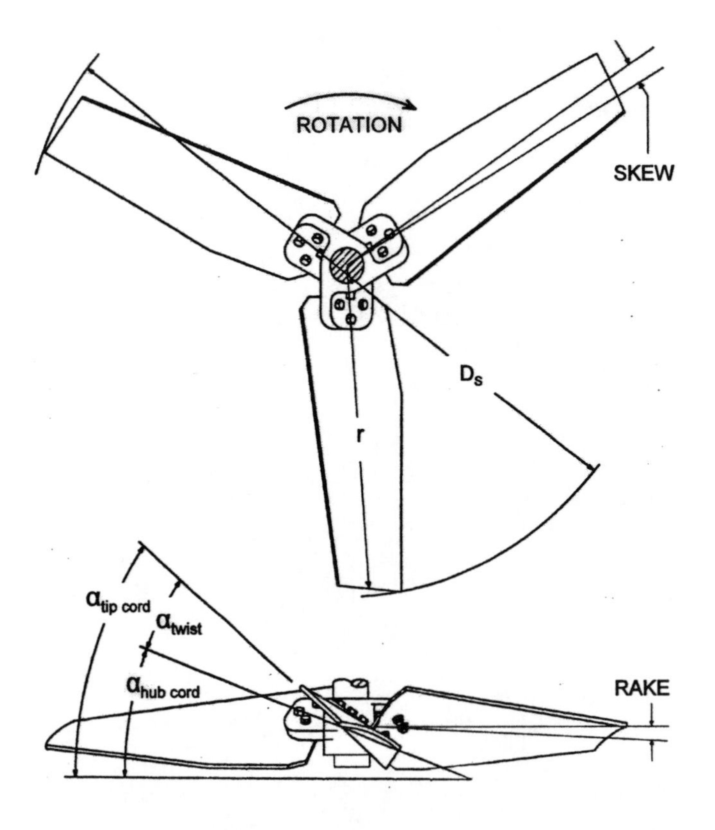

Fig. 202.10 Axial Flow Impeller

Because axial flow impellers can have many different shapes, no single set of nomenclature may be suitable for all designs. Different manufactures have different designations for similar designs. One should know the manufacturer's nomenclature as a first step in describing an impeller. The dimensions and nomenclature in this section reflect some more important characteristics and use designations similar to those found in naval architecture.

Impeller diameter, D_s, is usually taken to be the swept diameter of the impeller. However, twice the centerline radius, r, of the blade may also be used for impeller diameter. When the blade is rolled or bent, and the axis of the roll or bend does not parallel a radial line along the blade, an apparent twist in the blade results. Usually the angle at the blade tip is less than the angle at the hub. Typically, one should measure or know the blade angle at the tip and the angle at the hub attachment. The angle is usually measured from the horizontal to the cord (or a line connecting the leading and trailing edge of the blade). See Fig. 202.11 for detail of the cord angle. The twist angle is the difference between the cord angle at the hub and the cord angle at the tip.

The blade shape, curvature, and attachment may cause other differences in the apparent blade angle. If the radial line through the middle of the blade tip lies behind (in the direction of motion) the radial line through the center of blade attachment, the blade has a positive skew. The skew distance can be either positive or negative. If the line from the center of the hub through the middle of the blade tip lies below the horizontal, the blade has a positive rake. Both rake and skew may affect the discharge flow profile from an impeller.

For the purposes of fabrication, the shape of the blade cross section is usually defined in by roll radius or bend angle. With respect to performance, the effective airfoil shape is important. The airfoil cross section of a typical rolled blade is shown in Fig. 202.11.

The cord or developed blade width, W_d, is the distance between points on the leading and trailing edge of the blade that are at the same radial distance from the axis of rotation. The cord angle is measured from the horizontal. The projected width, W_p, of the blade is measured to the vertical. The height of the blade from the cord, h_{cord}, is the maximum distance from the cord to the blade. Blade thickness should be included in this height. If the roll is not uniform, or a bend or bends are used to approximate curvature, the maximum height may not be at the center of the cord length. The expanded width, W_e, of the blade is the width of the original piece of material used to form the blade. All these dimensions may change at different radii from the impeller hub.

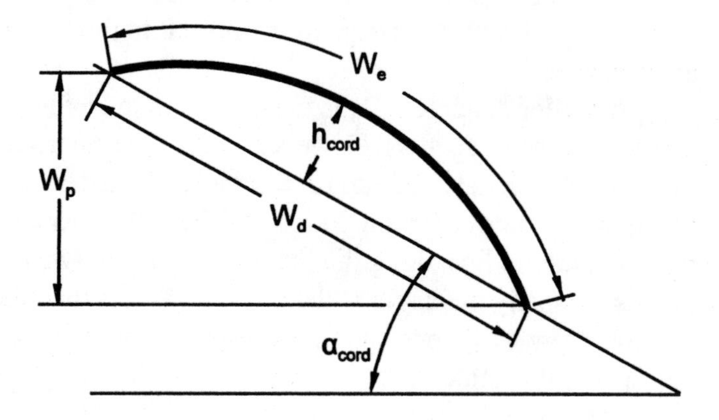

Fig. 202.11 Blade Cross Section

Blade areas can also be calculated based on projected, developed, or expanded width. When projected area, corrected for overlapping blades and the additional area of the impeller hub, is compared with the swept area of the rotating impeller, an impeller solidity can be calculated. Solidity is often used as a general quality describing blade width and number of blades.

Beyond the manufacturer's designation of the impeller, number of blades, impeller diameter, hub angle, developed width at the tip, and cord angle at the tip should be recorded as a minimum. Details of blade shape, roll radius or bend angle(s), rake, and skew are important primarily for fabrication. Most manufactures try to hold geometric similarity within an impeller type to be able to estimate power and performance.

202.2.5 **Marine Propeller** - A marine propeller can be used for mixing. Current use of marine propellers on mixers is limited by cost and weight to sizes less than about 12 inches (300 mm) in diameter. Most marine propellers used for mixing have three blades and a pitch to diameter ratio of 1.0 or 1.5. Pitch is the theoretical distance advanced by the blades in a single rotation of the propeller. The theoretical path followed by the blade is a helix. The nomenclature for propellers and proprietary axial-flow impellers may be taken from naval architecture. Ref. 808.8.

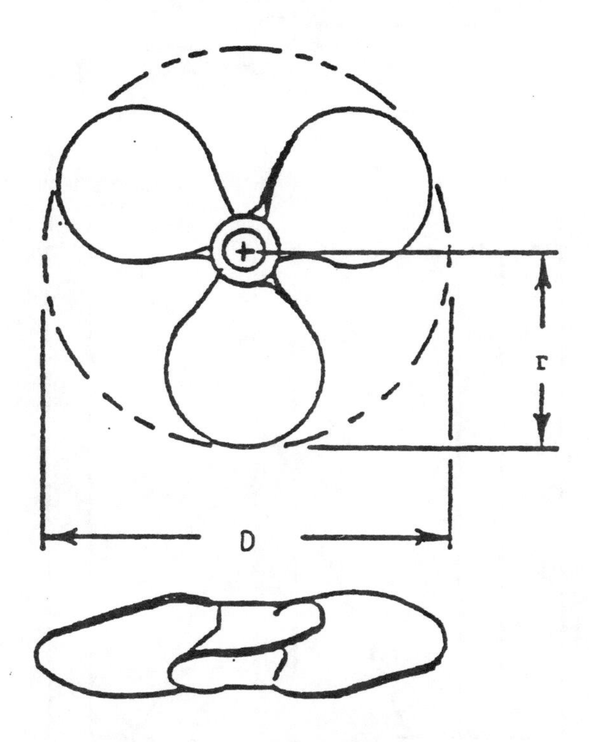

p = theoretical helical pitch

Fig. 202.12 Marine Propeller

202.2.6 **Anchor Impeller** - An anchor impeller is used primarily because the vertical arms pass close to the wall of the vessel, thus limiting how much material can remain motionless near the wall. To enhance this capability, flexible scrapers may be attached to the vertical arms. Anchor impellers and other close-clearance impellers are typically used in viscous materials that do not flow well. Close-clearance impellers rotate at slower speeds than turbines and other liquid pumping impellers.

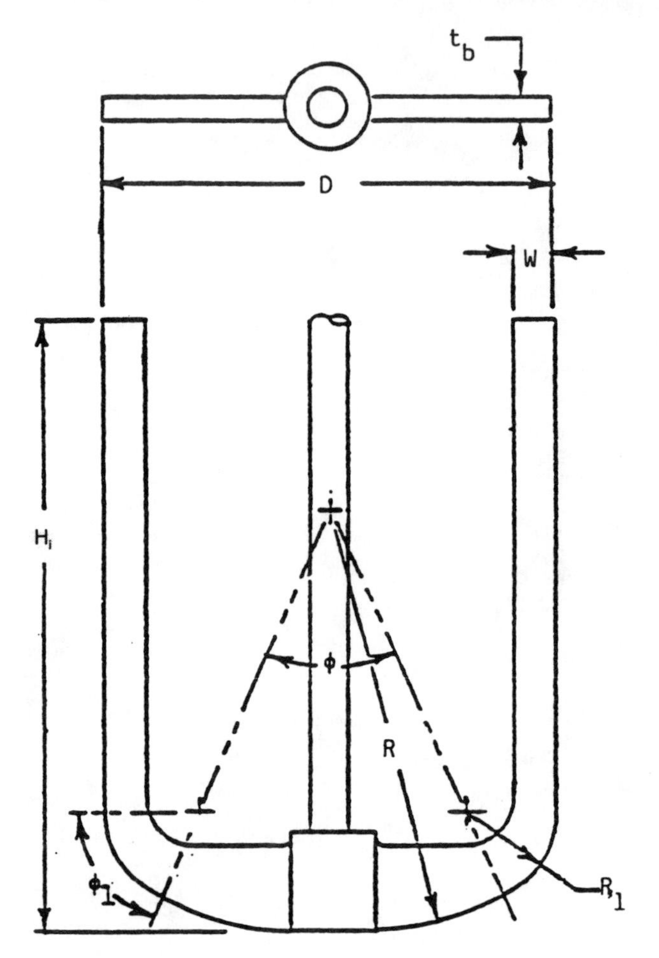

Fig. 202.13 Anchor Impeller

202.2.7 **Helical-ribbon Impellers** - Helical-ribbon impellers, like the double-flight helix shown in Fig. 202.14, operate with a close clearance to the tank wall like the anchor impeller. Helical-ribbon impellers cause better top-to-bottom motion than an anchor, which results in better mixing. However, a helix is more difficult to build and more expensive than an anchor. Helical-ribbon impellers may have one or two flights (continuous blade strips), may or may not have a sweep blade across the bottom, and can have different pitches.

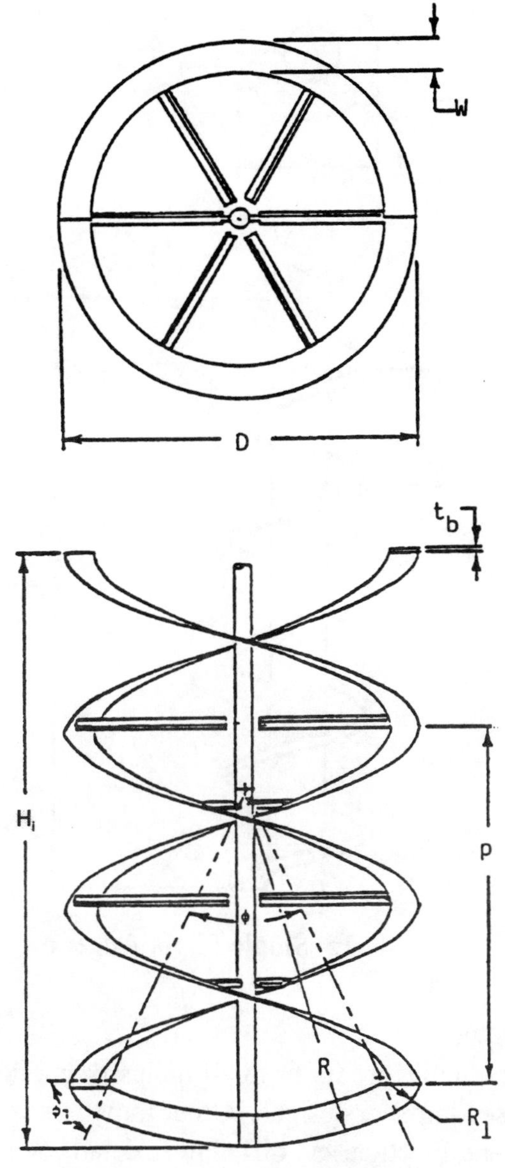

Fig. 202.14 Double Flight Helix

202.2.8 **Helical-Screw Impeller** - An auger or screw impeller, like the single-flight screw, shown in Fig. 202.15, is usually used in combination with other elements. Some screw-type impellers are used in a draft-tube (a stationary cylinder mounted inside the tank). Other helical screws are attached to the shaft of a helical-ribbon impeller to cause some pumping action in the opposite direction near the shaft.

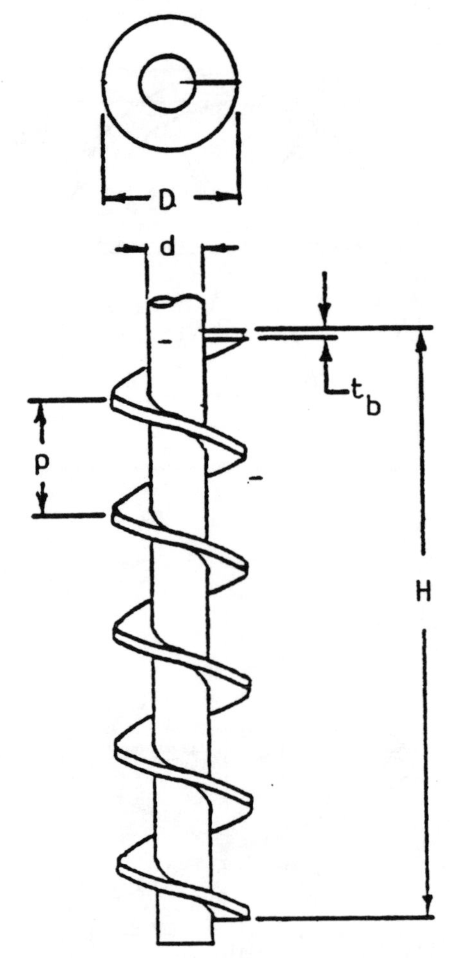

Fig. 202.15 Single Flight Auger or Screw

These examples are far from all-inclusive but do reflect many commonly used impeller designs. Other impellers are similar in construction and function, but different in detail. Some impellers incorporate features from more than one of the types shown. For additional information see Ref. 808.7, 808.9,808.12 and 808.17.

202.3 ***Vessels***

An impeller mixer is normally operated in a vertical cylindrical tank. However, horizontal cylindrical tanks are used occasionally. The cylindrical tank is used because of ease of fabrication from metals and the convenience in use. Tanks with square or rectangular cross sections are used when the material of construction is concrete. All length dimensions can be chosen to define a variety of both sizes and shapes.

Besides the vessel itself, baffles at the wall, impeller locations, and many other devices, such as: dip pipes, sparge rings, coils for heat transfer, feed points, and withdrawal points, are all part of the equipment. See the Glossary, Sec. 802, for additional descriptions of these components.

202.4 ***Auxiliary Equipment***

Auxiliary equipment such as compressors for gas sparging, pumps for liquid recirculation, external heat exchangers, and similar devices associated with the mixed process must be considered. Such equipment may contribute substantially to the performance and/or behavior of the mixing equipment.

203.0 *Basic Nomenclature*

203.1 ***Equipment Variables***

Many dimensions and parameters are necessary to describe the geometry and the operational performance of mixing equipment completely. Some commonly used terms are described in this section. These symbols and definitions will be throughout this testing procedures. For a complete list of the nomenclature, see Sec. 803.0

203.1.1 **Impeller Diameter,** D. Diameter is either measured as the maximum diameter swept about the axis of rotation or the twice the radius measured from the center of rotation to the center of the tip of a blade. The method of measurement is usually a matter of convenience, but measurements must be accurate and should be made for each blade or pair of blades. Impeller diameter is the most important dimension for mixer testing and must be accurately measured and clearly defined.

203.1.2 **Rotational Speed,** N. Speed is normally measured in revolutions per unit time, such as revolutions per minute or revolutions per second.

203.1.3 **Tank Diameter**, *T*. Since most mixed tanks are cylindrical, diameter is an appropriate measure of tank size, especially the inside diameter.

203.1.4 **Liquid Level**, *H* or *Z*. Liquid level is usually measured from the deepest point, such as the bottom of a dished head.

203.1.5 **Power**, *P*. Impeller power is the energy input characteristic most directly related to process results, yet measuring it directly is difficult. Several different power measurements can be made, some of which include losses not related to the process results.

203.1.5.1 Impeller power - power consumed by an impeller moving through the fluid.

203.1.5.2 Shaft power - power consumed by all of the impellers on the mixer shaft. Shaft power for multiple impellers on a single shaft is not always the sum of the individual impeller powers. This difference may be significant when impellers are spaced less than an impeller diameter apart.

203.1.5.3 Brake power - power input at the mixer drive (gear reducer input), which includes drive and seal losses.

203.1.5.4 Electric power - the power measured as volts and amps, with or without a power factor. Depending on the measurement method, electric power may include different electrical losses.

203.1.5.5 Motor power - the nominal (nameplate) power of the motor. The actual power delivered to the mixer is usually some fraction of the motor power to avoid overloads during normal operation.

Impeller power requirements depend on all of the impeller dimensions and sometimes the tank dimensions and impeller location. The primary variables affecting power are impeller diameter and rotational speed, which must be measured accurately. Motor power may be used to describe equipment, but does not necessarily reflect the operating performance of an impeller.

203.1.6 **Torque**, τ. Torque is related to power and speed by the relationship, $\tau = P/(2\pi N)$. Although torque is an important measure of equipment performance, it is even more important in determining the size of the mixer drive and mechanical strength of the shaft.

203.1.7 **Impeller Pumping Capacity**, *Q*. Pumping capacity, while not consistently defined, is often used to characterize the fluid motion resulting from impeller rotation. Primary pumping capacity normally describes the direct discharge from the impeller. Total pumping capacity may include some portion of the entrained flow, but definition must be provided.

203.1.8 **Fluid Density**, ρ. Density usually has a direct effect on impeller power under turbulent flow conditions. So a low-density hydrocarbon will require less power than water and a high-density solution or slurry will require more power. Density is often handled as **specific gravity** (*S.G.*), which is fluid density divided by water density.

203.1.9 **Viscosity**, μ. Viscosity is a measure of the resistance of a fluid to shear or flow. Consequently it is one of the most descriptive variables available to characterize liquids compared with how they may behave when mixed. The effect of viscosity is usually handled by the Reynolds number, which is defined later.

203.1.10 **Baffle Width**, W_B. Baffles are vertical obstructions used to prevent uncontrolled swirling of low viscosity fluid. Most baffles are just flat plates attached to the wall of a tank, but they may also be made of pipe, angle, or banks of coils.

203.1.11 **Impeller Clearance**, C. The clearance between an impeller and the bottom of the tank is an important variable in determining impeller position. Clearance is usually measured to the centerline of an impeller for positioning purposes. Clearance between the bottom edge of the impeller and the bottom of the tank may also be used. However, either clearance dimension must be checked for mechanical interference, especially in dished or conical bottom tanks. Whichever clearance dimension is used, it should be clearly defined.

203.1.12 **Shaft Length**, L. The shaft length is directly related to impeller clearance, location, and mounting configuration. Shaft length is one of the most important mechanical design variables. Again, this length dimension should be clearly defined and measured.

203.1.13 **Impeller Weight**, W_i. The weight of an impeller is important in shaft design, especially when calculating the shaft natural frequency.

203.1.14 **Blade Width**, W. The projected blade width is an important characteristic of the impeller blade, since it affects both power and flow.

203.1.15 **Blade Thickness,** t_b. Another characteristic of an impeller blade, primarily related to the strength of the blade.

203.2 ***Mixing Related Dimensionless Groups***

203.2.1 **Reynolds Number,**

$$Re = \frac{D^2 N \rho}{\mu}$$

The ratio of inertial forces to viscous forces.

203.2.2 **Power Number,**

$$Po = \frac{P g_c}{\rho N^3 D^5}$$

where g_c is the gravitational force constant, which converts from force to mass, for consistent units.

The ratio of imposed forces to inertial forces, which relates power to impeller and fluid characteristics.

203.2.3 **Pumping Number,**

$$Pu = \frac{Q}{N D^3}$$

The ratio of actual flow rate to a reference flow rate.

203.2.4 **Froude Number,**

$$Fr = \frac{N^2 D}{g}$$

The ratio of inertial force to gravity force.

203.2.5 **Aeration Number,**

$$Ae = \frac{Q_g}{N D^3}$$

where Q_g is gas flow rate.

The ratio of gas flow rate to a reference liquid flow rate. Volumetric gas flow rate depends on liquid head pressure. The gas flow rate most often used in mixing is based on the pressure and temperature conditions at the dispersing impeller location.

203.2.6 **Thrust Number,**

$$Th = \frac{F_{th} g_c}{\rho N^2 D^4}$$

where F_{th} is impeller thrust.

The ratio of imposed forces to inertial forces.

203.3 ***Process Related Dimensionless Groups***

203.3.1 **Weber Number,**

$$We = \frac{N^2 D^3 \rho}{\sigma g_c}$$

where σ is surface tension.

The ratio of inertial forces to surface tension forces.

203.3.2 **Nusselt Number,**

$$Nu = \frac{hD}{k}$$

The ratio of convective heat transfer rate to conductive heat transfer rate.

203.3.3 **Prandtl Number,**

$$Pr = \frac{c_p \mu}{k}$$

The ratio of momentum transfer rate to heat transfer rate.

203.3.4 **Blend Time Number,**

$$\Theta = t_{blend} N$$

where t_{blend} is the blend time. A numerical subscript on the blend time may be used to clearly define the degree of uniformity of a blend.

$$\Theta_{99} = t_{blend99} N$$

where 99 represents 99 percent uniformity.

Ratios of actual blend time to a reference time.

203.3.5 **Peclet Number,**

$$Pe = \frac{c_p D^2 N \rho}{k}$$

The ratio of momentum transfer rate to heat conduction rate.

203.3.6 **Sherwood Number,**

$$Sh = \frac{k_L D}{D_{AB}}$$

The ratio of convective mass transfer rate to diffusive mass transfer rate.

203.3.7 **Schmidt Number,**

$$Sc = \frac{\mu}{\rho D_{AB}}$$

The ratio of viscous momentum transfer rate to diffusive mass transfer rate.

203.3.8 Other dimensionless groups may be defined in terms of mixer dimensions or characteristics for specific purposes. Several additional dimensionless groups are necessary to define characteristics of multiphase or reactive systems.

204.0 *Operating Conditions*

Operation of mixing equipment includes directly related conditions, such as: mixer speed and power requirements; and indirectly related conditions, such as: feed rates, pressure, temperature, fluid properties, liquid levels, residence time, etc.

205.0 *Types of Tests*

205.1 *Operating Performance*

Mixer power, speed and torque are important indicators of operating performance, especially how close actual conditions are to those chosen by design. A quick check of these basic conditions could correct an installation or design problem before committing the system to process operation.

205.2 *Mechanical Conditions*

Many mechanical conditions must be met for a piece of rotating machinery to operate successfully. Beyond proper installation, conditions associated with alignment of couplings, shafts and gears, adjustment of shaft seals or drive belts, lubrication and general maintenance are all associated with equipment performance. Tests and measurements of mechanical conditions should always be considered before starting equipment, or when service records are in question.

205.3 *Mechanical Operation*

Simple observations, such as: direction of rotation and unusual noise or vibration may identify possible sources of problems.

Significant vibrations or shaft deflections are indications of serious mechanical problems and should be corrected immediately. Any loud noises or readily observed movements associated with the gear drive should be investigated.

205.3.1 Shaft deflections can be a result of or lead to serious mechanical problems, especially since most impeller mixing equipment is built with long overhung shafts.

205.3.2 Large shaft deflections are those where the amount of movement exceeds generally accepted engineering limits. Continued operation could result in premature bearing or seal failures or even catastrophic mixer shaft failure.

205.3.3 Large shaft deflections can result from simple mechanical causes or complex interactions between fluid forces and structural dynamics. The four most common causes are as follows:

205.3.3.1 **Mechanical** - bent shaft, unbalanced impeller, shaft not vertical, loose bearings or couplings, etc.

205.3.3.2 **Fluid** - strong disturbances, such as side flows in the impeller region, can cause imbalanced loads.

205.3.3.3 **Dynamic** - excitation of structural harmonics, especially related to frequencies associated with rotational speed or a multiple of the speed associated with an impeller blade passing frequency.

205.3.3.4 **Design** - structural defects or inadequate shaft diameter for even normal fluid or mechanical forces.

205.4 ***Mixer Support Structure***

The support on which the mixer is mounted is often overlooked as a design consideration. Whether mounted on beams or a nozzle, the mixer support must be stiff enough to handle a dynamic load. A deflection of an eighth of an inch (3 mm) may be perfectly acceptable for beam support of a static load. However, the same deflection with a dynamic mixer load may cause the end of the shaft to swing several inches to each side and the mixer will rock back and forth. The support structure should be stiff enough that hydraulic forces on the impellers will not cause the mixer shaft to move more than 1/32 inch per foot of length (2.6 mm per meter). A method for estimating hydraulic force can be found in Sec. 605.

Support stiffness is also a factor that can adversely affect natural frequency. Except for portable mixers, most mixers operate below the first lateral natural frequency of the mixer shaft. Impeller weight, shaft diameter, bearing spacing, and support stiffness all affect natural frequency. The natural frequency formula in Sec. 605 assumes that the bearing support is stiff. As support stiffness decreases, the natural frequency also decreases. A mixer on a weak support will operate closer to the natural frequency of the shaft and may cause destructive oscillations, not anticipated in the original design.

A typical open tank support structure is shown in Fig. 205.1.

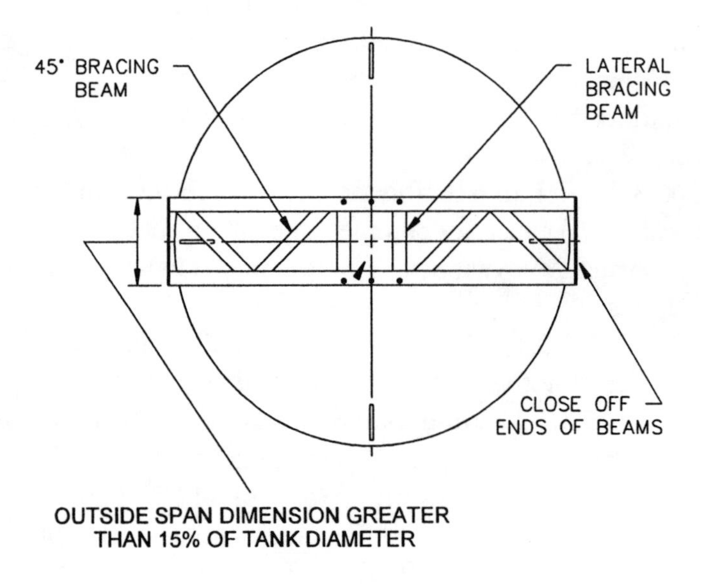

Fig. 205.1 Typical Open Tank Support Structure

The channels or beams in the support structure should be under the mixer drive as shown in Fig. 205.2. The support beams must be at the same level and the mounting surface flat to be sure that the mixer is vertical.

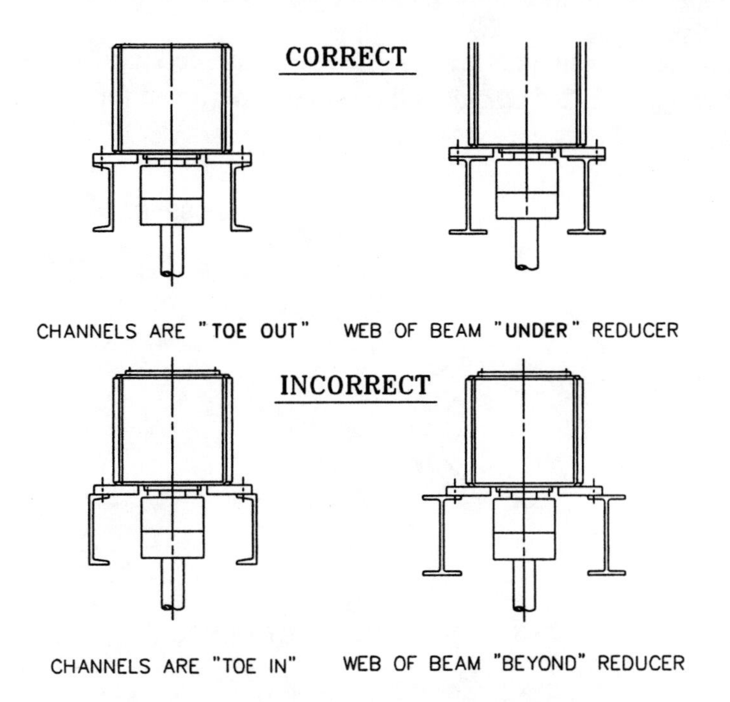

Fig. 205.2 Mounting on Beam Supports

Nozzles used to support mixers must be reinforced with gussets and pads as shown in Fig. 205.3, unless the vessel is designed for very high pressures.

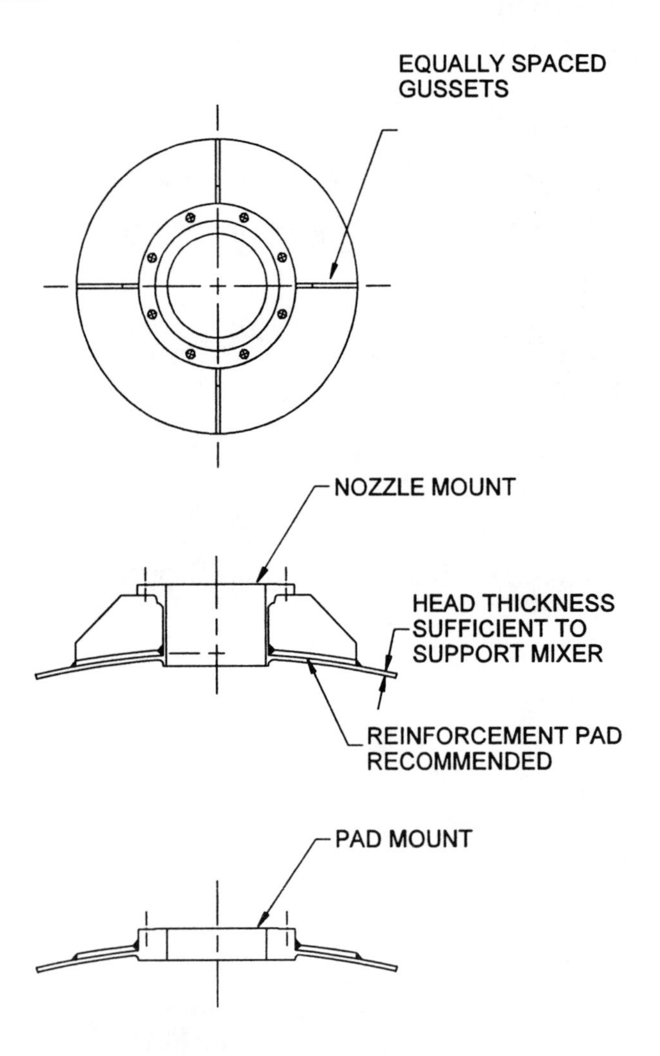

Fig. 205.3 Nozzle Gussets and Pads

Shaft straightness to within 0.005 inches per foot (0.42 mm per meter) and static balance are also necessary for smooth operation of a mixer.

205.5 ***Process Conditions***

Various types of mixing tests can be considered and used to evaluate many types of process performance. Tests could be run as part of an equipment certification procedure or to re-evaluate an existing piece of equipment. Such tests might include, for example:

205.5.1 *Miscible liquid blending* - to evaluate mixing performance such as blend time or degree of homogeneity when processing miscible fluids.
205.5.2 *Heat transfer* - to evaluate mixer performance including local or overall heat transfer coefficients from heat transfer surfaces.
205.5.3 *Immiscible liquid dispersion* - to evaluate mixer performance such as maximum or minimum droplet size, droplet size distribution, emulsion stability, and mass transfer when dispersing one fluid in another.
205.5.4 *Solids suspension in liquid* - to evaluate mixer performance such as level of suspension and distribution of suspended solids in a liquid.
205.5.5 *Gas-liquid dispersion* - to evaluate mixer performance such as gas hold-up, maximum bubble size, mass transfer rate, and reaction rate.
205.5.6 *Reaction rate* - may be influenced by several aspects of mixing, including uniform blending, heat transfer, mass transfer, and rapid or complete mixing may be reflected in the quantity or quality of reaction products.
205.5.7 *Variable conditions* - to evaluate mixer performance when conditions change, such as: liquid levels, phase ratios, viscosity, solids content, temperature, and pressure.
205.5.8 *Other tests* - because of the enormous variety of processes that use mixing equipment, other tests may have significance to a user.

206.0 *Performance Criteria*

Performance criteria should be established according to process results or intended purposes of the equipment. The performance of a mixer is limited by its original design and may not be suitable to achieve performance for which it was not intended. For example, a mixer designed to blend liquids may be unsuitable to suspend solids.

Performance of a mixer is more difficult to characterize than most types of process equipment because a mixer usually performs several functions at once. Mixer performance is often judged on whether it makes the correct or acceptable product, and not whether it functions hydraulically as designed. A failure in any aspect of the mixer performance may cause unsatisfactory results.

Determination of the performance criteria must focus on the most important result. The tests required to determine level of performance may involve many indirect measures of performance besides the specific results.

Typical mixer functions may include:

Batch Environment	Physical Processes	Transport Process
Liquid-Solid	Suspension	Dissolving
Liquid-Gas	Dispersion	Absorption
Immiscible Liquids	Emulsions	Extraction
Miscible Liquids	Blending	Reactions
Fluid Motion	Pumping	Heat Transfer

300.0 Test Planning

301.0 *Preliminary Considerations*

301.1 *Safety*

Any equipment testing must conform to the latest requirements of all applicable safety standards. These standards include, but are not limited to plant, industry, local, state, and federal regulations. It is recommended that all testing be conducted under the supervision of personnel fully experienced in plant and equipment operating practices.

301.2 *Environmental*

The test procedures must conform to the latest requirements of all applicable environmental standards. These standards include plant, but are not limited to plant, industry, local, state, and federal regulations. Environmental standards that apply to the equipment in normal operation should also apply during testing.

301.3 *Test Objectives*

The reasons for running a test must be clearly defined at the beginning.

301.3.1 Tests may be run to decide whether the mixer performs satisfactorily from a process and/or mechanical standpoint.

301.3.2 Process and mechanical requirements may be a contractual obligation with tests planned accordingly.

301.3.3 Tests may be conducted to identify a basis for scale-up or scale-down.

301.3.4 Tests may be run to identify possible causes of process or mechanical problems.

301.3.5 Tests may be part of an effort to improve process performance or productivity through equipment modifications.

301.4 *Multiple Applications*

Effects of changing liquid level, viscosity, speed, flow rates, and other process conditions may be important in tests. Some equipment is used for different purposes at different times. The testing requirements, may be different for different situations.

302.0 *Plans for Operating Performance Tests*

Power, torque, and speed are important measures of mixer performance, and indirect measures of process performance. A knowledge of these three variables is essential to testing of any impeller mixing equipment. In addition, the design of other vessel accessories and support structures are directly related to these variables.

Power data are also important for establishing operating costs, and often, are directly related to the process result.

To relate motor power to mixer power (energy applied to the process fluid), the efficiency of each component of the drive system needs to be defined. These efficiencies are often a complex function of operating speed and applied power.

302.1 *Speed*

Impeller speed should be measured to determine whether the equipment operates at the speed for which it is designed. Accurate measurement, typically to the nearest revolution per minute (hundredth of a revolution per second), is necessary because of the strong influence of speed on power.

The measurement is also required to relate power to torque. If the mixer is provided with variable speed capability, other tests may be done at maximum, minimum, and intermediate speeds to test the full range of operation.

302.2 *Power*

Motor power draw may be measured to check that motor nameplate rating is not exceeded and/or to decide if the impeller is imparting the design or desired power to the fluid.

302.3 *Torque*

Torque may be measured to check that the torque rating of the gear drive or other component is not exceeded, and/or to decide if the impeller is imparting the design torque to the fluid.

303.0 *Plans for Mechanical Condition Tests*

303.1 *Equipment Verification*

As an important part of the planning and preparation process, a thorough inspection and documentation of all aspects of the equipment should be done. A complete review of drawings, instruction manuals, rating plates, and any other documentation may help identify or correct anticipated problems.

303.2 ***Alignment***

Alignment of flexible and rigid couplings or belt drives may be measured to check if they are within design specifications. The gear drive may have to be checked to see whether it is level. Special adjustments, such as for a separate seal or steady bearing, may be required.

303.3 ***Runout***

Runout (lack of centering) of the impeller shaft and other shafts may be measured to check if they are within design specifications. Shaft runout is particularly important at a shaft seal or at the end of the overhung shaft.

303.4 ***Gear Tooth Contact***

Gear tooth contact patterns may be measured to check compliance with design specifications. A poor contact pattern can be an indication of mechanical defects.

303.5 ***Seals***

Most mixer shaft seals are designed to leak at a small finite rate for lubrication and cooling. Vessel contents (liquid or vapor) may leak out of the vessel with a single seal. With a double seal fluid may leak into the vessel. A double seal with pressurized barrier fluid must be used when the tank contents are hazardous chemicals. As part of the test procedure, the leakage rate may be measured, or the leakage analyzed for content, to decide if the seal is operating as designed.

For high temperature or high speed operation, a seal fluid recirculation system, with an external heat exchanger may be required to keep the seal at a safe operating temperature. A test of the seal system may be an appropriate part of the mixer test.

303.6 ***Auxiliary Equipment***

Tests may be required to ensure that auxiliary equipment is operating within design specifications. Auxiliary equipment can include: steady bearings, variable speed drive, lubrication system, cooling system, and other devices.

303.7 ***Vibration***

Vibration may be measured on the motor and/or the gear drive to decide if it meets design standards and to be sure that adverse vessel/support/mixing equipment interaction does not cause excessive vibration.

303.8 ***Noise***

Noise may be measured to assess compliance with design and/or environmental standards. Excessive noise may be an indication of a mechanical defect or adverse interaction with the vessel support system.

304.0 ***Plans for Mechanical Operation Tests***

Most mixers have cantilevered (overhung) shafts, i.e., the shaft is supported only at the drive end without a steady (foot) bearing at the bottom of the shaft. Lateral forces due to non-uniform flow act on the impeller to bend (deflect) the shaft. If the deflections are too large, seal or accessory life problems might be encountered. Ultimately complete failure of the shaft could also occur.

Before going further in planning a test, review the installation. The most common causes of shaft deflection problems are errors in mechanical installation, i.e., incorrect bolt torque, or damaged equipment. A complete dimensional check should be done first. Bent shafts, unbalanced impellers, out-of-plumb shafts, loose equipment, etc. are typical factors.

If a question about shaft deflections exists, the mixer designer/manufacturer should be contacted for specific information about acceptable limits. Dynamic tests should be considered only after problems have been documented. The testing procedure should be a sequential program to define and eliminate possible causes of large deflections.

304.1 Measure and quantify deflections. Identify runout and point of measurement.

304.2 Usually, any process factors that affect power input or liquid motion will probably affect fluid forces. Typical factors include: liquid level, gas sparging, liquid feed, baffles, accessories near the impeller, system asymmetries, etc. Because of this relationship, a review of power data could be helpful.

304.3 Measure natural frequency of shaft and mounting structures.

304.4 Measure shaft strain or deflection under operation. If possible, vary speed and measure changes in shaft strain or deflections.

304.5 Operating a mixer with an impeller near the liquid level may cause sever fluid forces in excess of design limits. Know whether the mixer was designed to handle the loads caused by liquid level changes. Do not operate the mixer while filling or emptying the tank, unless it was designed to handle such loads.

305.0 ***Plans for Process Condition Tests***

If multiple process variables are an essential part of a testing program, a systematic approach to experimental design is recommended. Only through experimental design can process effects be adequately decoupled for analysis and interpretation.

305.1 ***Miscible Liquid Blending***. Measure time required (blend time) to achieve a specified degree of uniformity. Sample volume and location must be specified, besides uniformity criteria to achieve consistent results. See Ref. 808.9, 808.7.2, 808.7.4, 808.7.7 and 808.7.8 for additional information about mixing.

305.2 ***Heat Transfer***. Overall heat transfer coefficients may be determined and an estimated process side heat transfer coefficient computed. See Ref. 808.9, 808.12, 808.7.5, and 808.24 for additional information about heat transfer.

305.3 ***Immiscible Liquid Dispersion***. Dispersion may be tested by physical criteria such as droplet size distribution or mean droplet size or by mass transfer criteria, i.e., mass transfer coefficient.

305.4 ***Solids Suspension in Liquid***. Usually percent solids and size distribution as a function of position in a vessel is measured. For incomplete suspensions, fillet size and contour may be measured. See Ref. 808.12, 808.17, 808.7.9 and 808.7.12 for additional information about solids suspension.

305.5 ***Gas-Liquid Dispersion***. The usual criteria in gas dispersion are related to mass transfer, e.g., k_La measured according to certain assumptions by standard or special methods.

305.6 ***Variable Conditions***. Changes in physical properties such as viscosity, or other operating conditions may influence test results for any of these process tests and should be taken into account.

305.7 ***Other Measurable Results***. Tests could be run to evaluate many other process parameters. Tests must be appropriately planned so that the needed evaluations can be made.

306.0 *Performance Criteria*

A precise numerical definition should be established, if possible, on each test to decide what is successful, or at least if improved performance has been achieved.

400.0 Measurement Methods & Instruments

401.0 *Introduction*

A variety of equipment is available for measuring the characteristics of the mixer and the process. The following sections describe some more important equipment and tests that can provide information necessary for performance tests.

In developing technical information it is important that the accuracy and the rigor of the development of the information match the required application.

The results need not be any more precise or accurate than required for the intended application. Sometimes a preliminary measurement using simple methods will establish whether the result is important and whether more accuracy is required. The calibration accuracy and data reproducibility (precision) must be defined for each measurement.

402.0 *Operating Performance Measurements*

402.1 *Speed*

The speed of the mixer shaft can be measured by using a tachometer, a stroboscope, or by counting the low speed shaft rotations. Any device should have an accuracy of plus or minus 1 percent of this important measurement. Uncertainty in power will be less than plus or minus 3 percent with this accuracy.

The use of nameplate motor speed and nominal gear ratio is not sufficiently accurate for speed determination. Motor slip and variance within nominal speed ratios, accepted by American Gear Manufacturers Association (AGMA) Standards, can contribute to significant errors compared with actual speed. Motor speed can be used only if it is accurately measured. Under a full load, an 1,800 rpm motor will typically turn at about 1,750 rpm, but could operate between 1,730 rpm and 1,770 rpm, which may cause errors in mixer observations and calculations. The exact gear ratio of the drive can be used if it is known or can be determined.

Do not rely on frequency output from a variable frequency drive controller as a measure of motor speed, especially if viscosity changes can affect loads. Shaft rotational speed is not a linear function of controller frequency under heavy load conditions.

402.2 *Power*

The measurement of power is an important characteristic of the mixer performance. Besides mixer design, vessel hardware and support structure are directly related to power requirements. Power data are also important for establishing operating costs and are usually related to process results.

402.2.1 Electrical Power Measurements

Direct measurement of electric power (kilowatts) is accomplished by using a wattmeter. This method is preferred for determining the power drawn from alternating current, induction motors. A good wattmeter contains the circuitry necessary to measure volts, amps and phase angle (power factor or voltage-amperage reactance), and thus to reflect true power accurately.

402.2.2 Other Electrical Measurements

Power may be calculated quite accurately if amperage, voltage, and power factor are measured. Accuracy decreases if power factor and/or voltage are not measured, errors can be greater than 20 percent.

402.2.2.1 Current

An ammeter is used to measure the electric current. Clamp-on induction coil meters are commonly used. For multiphase motors, the current needs to be measured for each active leg.

402.2.2.2 Voltage

A voltmeter is used to measure electric voltage. For multiphase circuits the voltage should be measured between active legs, and readings matched with current readings.

402.2.2.3 Power Factor

Although the power factor is frequently thought to be a characteristic of an individual circuit element, it can be strongly dependent on other equipment installed on the line. For the highest degree of accuracy, the power factor should be measured for each application. A low power factor may even be a reason for motor overloads.

402.2.3 Component Losses/Efficiency

A complete mixer analysis includes the performance of several drive components: motor, couplings, variable speed drives, gear reducers, etc. Power is often not identified by the point of measurement.

The difference between input power (usually kilowatts) to an electric motor and the power applied to the fluid (shaft power) may exceed 15 percent. Principal losses include variable frequency drives, motor efficiency, gear reducer losses, and seal friction. Converting a measurement of power from one point to another requires a detailed knowledge of the power losses for each component. The loss (or efficiency) for several components depends on both operating speed and transmitted power.

402.2.4 **No-load Power**

The full-load drive losses cannot be accurately measured by operating the mixer in air, a "no-load" condition. Besides measurement accuracy, component losses at no-load conditions are often different from losses at full-load conditions. Gear losses will increase with load. Measurement of no-load losses may sometimes suggest the actual losses or identify other problems.

402.2.5 **Other Power Measurements**

Since power sources other than electricity can be used to drive a mixer, other measurements may be appropriate for a given application. For example, power can be determined for air motors and hydraulic drives from measurements of pressure drop across the motor and the operating speed (flow rate).

402.3 *Torque*

Direct measurement of torque is most often used as an alternative method of determining power requirements. The product of torque and speed for a rotating shaft is a direct representation of transmitted power. The accuracy of power measured by these techniques depends on cumulative errors in the individual measurements. The careful application of torque measurement techniques can yield an accurate and direct measure of impeller power.

402.3.1 **Rotating Shaft**

Instrumentation is available to measure the torsional loading on a rotating shaft. These devices are generally based on a flexural member with a strain gauge bridge attached. Associated circuitry powers the bridge and measures the resistance imbalance due to torsional loading. These devices may be subject to error induced when the strain gauge region is subjected to excessive bending loads. Errors can also be introduced during the transmission of the signal from the rotating shaft to the observation point.

402.3.2 **Reaction Load**

On small scale tests, the vessel or drive system can be mounted on a low friction bearing. The torque required to prevent rotation of either the vessel or drive is measured. This technique has limited applicability in large scale equipment.

402.3.3 **Reaction Strain**

Torque might also be measured by the deflection of motor mounts or other support structures. Materials loaded in the elastic range will undergo deflections directly proportional to the associated load. A suitably precise measurement using strain gauge techniques to measure these deflections can be used to measure torque.

402.3.4 **Calibration**

For any custom measuring technique, a detailed calibration must be done. Tests must be devised to quantify the precision and the accuracy of the measurements. The most convenient method of calibration involves applying known static loads. The dynamic mixer loads can be compared with the static calibration. The calibration must include any effects of installation geometry that might affect the measurements.

403.0 *Mechanical Condition Measurements*

If possible, measure all dimensions of the installed equipment and keep dated records. Photographs of the equipment and tank internals also provide an excellent permanent record for subsequent comparison with future conditions.

403.1 *Alignment and Adjustment*

403.1.1 **Couplings and Belt Drives**

Alignment of flexible couplings, rigid couplings and belt drives is normally accomplished with a scale and a clamp-on or magnetic-base adjustable-arm dial indicating micrometer or other instruments as specified by the manufacturer. Electronic indicators are also available. Errors in coupling or belt alignment can result in high-frequency vibrations.

403.1.2 **Gear Drive Base**

A level and feeler gauge can be used to check whether the gear drive is level to within the specifications of the manufacturer.

403.1.3 **Vertical Alignment**

Proper leveling and settling of the gear drive normally satisfies vertical alignment, however, the vertical alignment should be checked, and may be used to check drive leveling.

403.1.4 **Special Requirements**

Manufacturers may say special adjustments are required.

Refer to the following sections for separately mounted seals and steady bearings.

403.2 ***Runout***

Shaft runout may be measured with a magnetic-base dial indicating micrometer, displacement proximity probe, or other suitable device. A measurement range of 0-50 mils (0-1000 microns) peak-to-peak displacement with 0.5 mils (10 microns) graduations would normally be sufficient for measurement at the drive base or seal.

403.3 ***Gear Tooth Contact Pattern***

Gear tooth contact patterns may be measured with marking compound, commonly termed transfer dye or bluing dye. A spray coating of molybdenum disulfide may also be used to mark the gearing. Depending on the type of gearing, the exact pattern may be different, but usually it should show consistent and uniform contact on all the teeth.

403.4 ***Seals***

403.4.1 **Packing Seals**

A packing seal usually consists of a pliable compound or rope held tightly around the shaft. Although this method of sealing can retain a considerable pressure, a certain amount of liquid or lubricant leakage is normally necessary to reduce friction and remove heat generated by friction. The rate of leakage may be important. Some packing compounds, especially those with a graphite or graphite and PTFE base, may not require grease lubrication.

403.4.1.1 Liquid leakage out of a submerged packing seal (stuffing box seal) may be detected visually. Liquid may be collected over a period and the volumetric leakage rate determined.

403.4.1.2 Grease consumption may be the only leakage in or out of a packed seal located above the liquid level. Grease requirements over a period of time, up to weeks or more, may be logged and reported as a volumetric consumption rate.

403.4.1.3 Vapor leakage out of a packed seal may be detected with soapy water. A leakage rate may be recorded as bubbles per minute. The vapor may be collected if a backup seal is provided and analyzed for content and rate.

403.4.2 ***Mechanical Seals***

A mechanical seal with a liquid barrier lubricant is the seal type least likely to show visible leakage, although lubricant may leak in to or out of the vessel. Lubricant leaking out of the seal may vaporize and not be visible. The volume of lubricant in the barrier liquid reservoir may be measured periodically and additions logged for weeks or months to be reported as a volumetric consumption rate. The temperature of the barrier lubricant may be measured at operating conditions using a suitable thermometer.

403.4.3 **Separately Mounted Seals**

Most seals will be mounted in a rigid frame attached to the gear drive, which ensures alignment with the mixer shaft. However, some types of seals, typically called "separately mounted seals," must be aligned after installation. Concentricity and squareness of the seal to the shaft may be measured with a clamp-on or magnetic-base dial indicating micrometer. More accurate alignments can be made with laser alignment systems.

403.5 ***Auxiliary Equipment***

403.5.1 **Steady Bearings**

If needed, a steady bearing is usually installed after the gear drive and shaft have been fixed in place. The steady bearing housing can be set in place to suit the shaft using a rule and scribe.

403.5.2 **Lubrication and Cooling Systems**

Pressure measurements may be made at various locations with a pressure gauge. Pressure gauges may be installed permanently.

Flow rate for proper operation may be verified directly by flow measurement or indirectly by temperature measurement.

403.6 ***Vibration***

Vibration of equipment components may be measured with the following types of instruments:

403.6.1 a contact-type displacement transducer suitable for 0-10 mils (0-250 microns) peak-to-peak displacement with 0.2 mils (5 microns) graduations and a filter for 1-100 Hz.

403.6.2 a contact-type velocity transducer suitable for 0-1 inches/second (0-25 mm/s) peak velocity with 0.02 inches/second (0.5 mm/s) graduations and a filter for 1-100 Hz.

403.6.3 a contact-type accelerometer suitable for 0-5 g's (0-50 m/s^2) peak acceleration with 0.1 g's (1 m/s^2) graduations and a filter for 1-100 Hz.

403.7 ***Noise***

Field noise measurements would normally be for sound pressure level in decibels (dB). The following types of instruments may be used:

403.7.1 Sound level meter with an A-weighted filter network suitable for 50-120 dBA with 1 dBA graduations.

403.7.2 Octave band analyzer suitable for 50-120 dB with 1 dB graduations with standard set of contiguous octave bands in the range 60-8000 Hz.

403.8 ***Temperature***

The temperature of the oil in a gear drive or motor surface temperature may be measured with an indicating device suitable for 0-15 C with 1 C graduations.

404.0 *Mechanical Operation Measurements*

404.1 ***Natural Frequency***

Shaft and structure natural frequencies are measured using vibration instrumentation. Special low frequency resolution probably will be required. Large mixer shaft frequencies are often between 0.3 and 2.0 Hz.

404.2 ***Shaft Strain***

Stress is directly proportional to the strain as measured by foil gauges. A strain gauge bridge may be mounted on the shaft or a separate spool piece inserted between the mixer and drive. Signal amplitude increases as the location point is moved closer to the bearing next to the overhung portion of the shaft.

Signals must be transmitted from the rotating shaft to the stationary surroundings. Slip rings or radio telemetry can be used. Dynamic stress can be recorded on a strip chart recorder to get peak values. A spectral analysis of this signal gives pertinent frequency information. Frequencies may help identify the source of problems.

The output from these gauges can be used for natural frequency measurements in place of the acceleration probe.

404.3 ***Deflection***

Shaft deflection is a primary indicator of shaft dynamics. Dial indicators, magnetic proximity probe, and strain gauge deflection arms (i.e., "flipper gauges") have been used to measure deflections. Usually, these devices are mounted above the liquid level.

Deflection magnitude increases as the distance from the bearing increases. Greatest resolution would be obtained by having a sensor near the impeller on a cantilevered shaft.

A limitation for any of the direct shaft deflection measurements is that the static runout must be subtracted from it. It is conceivable that at the point of measurement, equipment runout (though within acceptable tolerances) is a significant part of the total deflection.

With suitable placement and signal processing, a displacement sensor could be used for natural frequency measurements in place of an acceleration probe.

404.4 *Spectral Analysis*

The shaft strain or deflection under operating conditions gives magnitude and frequency data. The ability to do a spectral analysis of these signals is a powerful aid in tracking down the source or cause of mechanical vibrations. Besides natural frequencies of shafts and structural components, gear mesh and bearing contact frequencies can be measured.

405.0 *Process Condition Measurements*

405.1 *Density or Specific Gravity*

Mixing tests will usually involve the measurement of fluid density, since power consumption is directly proportional to density except at low Reynolds numbers. Other fluid forces on the mixer are also proportional to density. Fluid blending applications frequently involve the blending of fluids of different densities. Apparent fluid density can be affected by entrained gases and suspended solids.

Density can be readily measured for homogeneous fluids. Hydrometers may be used, when available. An accuracy of ±0.01 can usually be obtained with such measurement techniques. A known volume can be weighed in a graduated cylinder, usually with a little less accuracy, but with satisfactory results.

For nonhomogeneous fluids, a variety of methods are available, including those just mentioned, to obtain a pseudo homogeneous value.

Average densities of liquid-gas systems are usually estimated by determining gas hold-up. The hold-up can be determined by the volume change on settling or by other methods, such as pulse testing. Measurement of the liquid surface may be difficult because of foam.

Densities of liquid-liquid and liquid-solid mixtures may be determined by calculating densities of the various phases and determining the volume fraction by settling or other separation techniques.

The average density (ρ) of a mixture can usually be estimated by

$$\rho = \left[\frac{x_1}{\rho_1} + \frac{x_2}{\rho_2} + \ldots \frac{x_n}{\rho_n} \right]^{-1}$$

where x_i and ρ_i are the mass fractions and densities respectively of each component. Average densities should not be calculated in this manner if a significant volume change on mixing occurs.

405.2 ***Viscosity***

Viscosity is an important physical parameter in many mixing tests. At viscosities less than 100 cp (0.1 Pa·s), viscous effects are typically small. At higher viscosities, the viscosity determines how difficult the movement or flow of the fluid will be. At viscosities greater then 10,000 cp (10 Pa·s), the effects may dominate the problem.

Process effects are usually correlated by a Reynolds number, and as such, related to many other parameters, such as power number, impeller pumping number, and heat transfer coefficient. In the laminar region of impeller Reynolds number, power consumption is proportional to viscosity, rather than density.

As much as possible, viscosity should be measured at the operating conditions of the mixer, including temperature, pressure, and shear rate. The viscosity of a Newtonian fluid is relative easy to measure, because the viscosity does not change with shear rate. Non-Newtonian fluids exhibit different viscosities at different shear rates and are more difficult to measure. Several types of viscometers can be used, including: open spindle, falling ball, cup, and orifice, close clearance cup and bob, or cone and plate, each with different shear characteristics and viscosity limits.

Shear stress may significantly change the apparent viscosity of some fluids, which may greatly affect mixer performance. Non-Newtonian behavior of fluids can be estimated by operating a viscometer at varying shear rates. Such tests could include running an open spindle viscometer at several speeds. A discussion of non-Newtonian fluids can be found in Ref. 808.9, 808.12, and 808.7.8.

Viscometers are available to measure viscosity using a spindle shaped much like a mixer paddle. Laboratory mixers with torque measurement capabilities may also be used. Operation at several speeds must be used to find viscosity at different shear rates. Slurry viscosities are best measured in some type of well mixed viscometer, since liquid and solids may separate or settle, resulting in erroneous viscosity measurements.

The viscosities of some fluids change with time. The tests must measure viscosity of the fluid sheared at a known shear rate for a period of time. Other fluids have different types of shear dependent behavior, including viscoelasticity.

405.3 *Physical State*

405.3.1 **Gas-Liquid and Liquid-Liquid**

Hold-up is the amount of dispersed phase maintained in a mixed system, and is usually expressed as a volume fraction or percent. The dispersed phase can be either a gas or liquid. Hold-up is frequently measured by collecting a sample of the material, allowing it to settle, and measuring how much material is in each phase.

In a flow through system, hold-up does not necessarily equal the volume fraction of the added dispersed phase. If density differences exist, one phase may travel faster than the other.

Drop size of a dispersed phase depends on the component properties and the equipment design and operation. Drop sizes can be measured by such techniques as photography and light transmittance.

405.3.2 **Solid-Liquid**

Particle size distributions can be reported in many ways. For sieve analyses, the results are usually reported as the weight fraction that goes through one screen and is caught on the next, such as -48 +65, for particles that pass through 48 mesh and are caught on 65 mesh screens. Results are also reported as cumulative mass fractions of diameters larger or smaller than a given size.

Solids settling velocities are needed for proper design of mixers and thus may be needed to run an accurate mixer test. Higher settling velocities require higher fluid velocities for suspension. For dilute (<10 percent), near-spherical particles, settling velocities can be measured by timing the fall of a sample of solids through a known distance in the test liquid.

405.4 *Chemical Composition*

Sometimes knowing the chemical composition of the fluid is necessary. Direct determinations can be made, such as by chemical analyses and various spectroscopic means.

If measuring composition or concentration during an operational test is necessary, it is more common to determine chemical composition by indirect means. Some physical or chemical property is selected that will vary with the change in composition. Appropriate calibrations are needed to decide the composition associated with the magnitude of the property. The property selected will depend on the fluid being evaluated. Tests to be used might include: density, refractive index, pH, light absorption, and even on-line chromatographic separations. Care must be taken to select a property that is not affected by other system changes and that varies enough so that experimental error does not significantly affect the results.

405.5 *Other Physical/Chemical Properties*

Many physical or chemical property measurements are possible. Sometimes, any of the following parameters might be important: Interfacial tension or surface tension, specific heat, thermal conductivity, boiling point, melting point, latent heat of vaporization or fusion, heat of reaction, heat of mixing, molecular weight, or mass diffusivity. Frequently, measurement is not required, because properties may be estimated from tabulated or empirical data. Even product quality may be a satisfactory measure of mixer performance or change in mixer performance.

406.0 *System Operating Conditions*

406.1 *Temperature*

Process temperature can affect the process directly and could change what product is made. Temperature changes can also relate to the performance of mechanical components, such as in seals or gear box oil.

A variety of temperature measuring, recording and controlling devices are available. The choice will depend on the specific requirements. For specific suggestions and measurement techniques, see Ref. 808.2.

Calibrating all temperature measuring devices before and after each test is desirable. Calibrated spares are often useful during a test.

The location of the measuring device is very important. It should be installed to minimize the error introduced by radiation or convection, and should be located in moving and mixed streams to obtain maximum sensitivity and to avoid errors due to stratification or stagnation.

406.2 *Pressure*

Equipment tests may be needed at elevated pressures, such as tests for seal leaks. The volume of a gas present in a gas-liquid test will vary with pressure.

Many types of pressure measuring devices are available commercially. The choice will depend on specific requirements. See Reference 808.1 for specific suggestions and measurement techniques.

Sometimes the process fluid needs to be isolated from the measurement element in the pressure gauge to prevent damaging, plugging, solidification, etc. Options include liquid seals or integral diaphragms to isolate the instrument from the process fluids.

406.3 *Liquid Level, Volume, and Flow*

406.3.1 **Liquid Level**

The type of instrument used to measure fluid level depends on the equipment size, the conditions of the test and the accuracy needed.

Gas hold-up for gas-liquid processes can cause the liquid level to change significantly from gassed to ungassed conditions.

406.3.2 **Volume Measurement**

Differential volume and differential weight are commonly used to measure flow of fluids from tanks. When fluid flow measurement is made using volumetric or weigh tanks, several measurements should be made to ensure that steady state is obtained before the start of the test.

Volumetric tanks should be calibrated before a test with weighed increments of liquid measured at a known temperature. Suitably designed volumetric tanks can be accurate within ±0.5 percent of total volume. Weigh tanks should be calibrated for volume over the entire range of weights at which they are to be used. Accuracies of ±0.2 percent of full tank weight can be obtained. Recommendations as to the proper design, construction, calibration, and operation of volumetric and weigh tanks are given in Ref. 808.3.

406.3.3 **Flow Measurement**

406.3.3.1 **Volumetric Meters**

The selection of meter type depends on the fluids being considered and the accuracy needed. However, the meters listed here measure volume and not mass flow. Errors will be caused by any variations in actual specific gravity, such as those caused by temperature fluctuations, entrained gas, etc.

Rotameters of many types are commercially available and, when properly installed, calibrated and cleaned, are excellent devices for fluid-flow measurement. Accuracies of 0.5 to 10 percent of full scale are common, depending on size and calibration. For information concerning corrections for temperature and pressure, see Ref. 808.3.

The rate of flow of fluids is most often measured by differential pressure meters. Commonly used ones are orifice meters, venturi meters, flow nozzle meters, and pitot tube meters. Accuracies of 0.5 to 10 percent of full scale are common depending on the type of meter used. For additional information, see Ref. 808.3.

Magnetic flow meters are suitable for liquids that have slight electrical conductivity. This style meter is reported to be particularly useful for measuring flow of liquids containing suspended solids. Accuracies of about 0.5 to 2 percent of full scale are common.

406.3.3.2 **Mass Flow Meters**

This style of meter is designed to handle multiphase fluids or applications where the specific gravity is not constant. Different styles can handle dispersed gases or solid slurries. Accuracies of 0.4 percent of actual rates are possible with some models.

406.3.3.3 **Weirs**

Weirs may be used for fluid-flow measurement of large flows, such as flow through large basins. Accuracy of 2 to 5 percent of full scale may be obtained with a calibrated weir.

406.3.3.4 **Impeller Flow Measurement**

Flow produced by an mixer impeller can be a significant item. Various laboratory methods have been developed to measure flow including pitot tubes, hot wire or hot film anemometers, laser anemometers, rotary impeller flow meters, etc. For most of these methods measuring the velocity at multiple points and then summing the values over an area is necessary. One has to be careful, however, to distinguish between the primary flow produced by an impeller and the total flow, which includes induced flow in the vessel. Impeller flow also has various vector components, which include axial (along the shaft), radial (out from the shaft), and tangential (in the direction of shaft rotation). Actual experimental measurements are difficult in large installed equipment. Visual observation may show surface motion, but little else. Impeller flow can be estimated using the data given in Sec. 804.2, although this is only an estimate.

Few process results are directly correlated to impeller flow. Process results, such as blend time and solids suspension, exhibit relationships involving multiple variables, including impeller type, impeller-to-tank diameter ratio, off-bottom clearance, and other factors. Impeller flow is important to the extent that the entire contents of a tank need to be in motion for most successful operations.

406.4 *Phase Ratios*

A determination is sometimes needed for the phase ratio of multiphase fluids. This can include liquid-solid, liquid-liquid, liquid-gas, and combinations.

Phase ratios are normally determined by obtaining a representative sample and then allowing the phases to separate, either by gravity or by centrifugal force. The amount of each phase must then be measured.

Phase ratios are generally expressed on a volume basis at specific conditions. Some devices are available that may directly or indirectly measure the composition of the phases without first separating them.

Determining the composition of the various phases may be necessary due to possible solubilities of different components in the phases.

406.5 *Blend Time*

Blend time can be defined as the time from the start of mixing at some unmixed condition until the vessel contents reach a predetermined value of uniformity. Frequently used criteria include the time to reach specified variations in temperature, density, component concentration, etc.

No universally accepted definition exists of what constitutes complete blending. Some processes require as little as 95 percent uniformity, while others may require more than 99.9 percent. Methods of determining uniformity must be considered. The degree of uniformity must be established based on process objectives or decided for each specific case. Since concentration varies with location, a statistically significant number of samples and locations are required to assure uniformity.

406.6 *Sampling*

Different types of sampling techniques can be used, depending on what is being measured.

Caution: grab-sample techniques should not be attempted while equipment is operating.

Some sampling methods include:

406.6.1 Grab scoops, bottles, buckets, etc., for surface samples.

406.6.2 Spigots and sample cocks for samples along walls.

406.6.3 Sample valves on discharge lines for exit condition samples.

406.6.4 Sampling pumps with movable sample lines for composite samples or sample from various locations.

406.6.5 A drop-bottle with a top that can be opened at the bottom or other known location may be used for multiple samples to estimate the average composition.

406.6.6 Samples should be taken from those points where the most useful data will be gathered rather than from those points where obtaining it is easiest. Depending on the reason for the test, one may need to sample the central region, quiet zones away from the mixer, the surface, the bottom, corners, etc.

406.6.7 Multiple samples should be taken to find statistical significance. Samples should be checked during the test to decide whether the sampling is sufficient as far as number, size, location, etc. Samples should be taken from more than one point depending on the degree of statistical certainty needed. Acceptable statistical methods should be used to find means and deviations, and the reliability of each. This should be done to define size and number of samples and to check the sample variance, Ref. 808.11 provides additional information on sampling.

500.0 Test Procedures

501.0 *General Test Procedures*

Impeller-type mixing equipment includes a broad variety of specific types of equipment, which may be applied to an even greater variety of processes. To define a general procedure that would be applicable in every case is impossible.

The best advice for testing any complicated combination of process and equipment variables is to first define the problems and then determine the most likely approaches to solving them.

The complexity of most processes compared with the performance of mixing equipment is such that only indirect measures of equipment performance are practical. As a result, many of the following procedures emphasize generally applicable measurement techniques that are often indicative of process performance or results.

502.0 *Preliminary Operation and Safety*

Before operating any mixing equipment, a thorough safety check should be done, including at least the following steps:

502.1 *Instruction Manuals*

Check the manufacturer's instruction, operation, and maintenance manuals for start-up procedures. Check for recommended oil and grease quantities, levels, and specifications.

502.2 *Check for Debris in Vessel*

The mixing vessel should be checked for debris, workmen's tools, ladders, and the like.

WARNING: Personnel should never be in a mixing vessel when the mixer is running, whether the vessel is empty or not. The mixer should be locked-out before anyone enters the vessel.

502.3 ***Check for Obstructions***

The mixing vessel should be checked for internal obstructions to the rotation of the impeller. Baffles or probes may have been installed inadvertently such that they would interfere with the impeller when rotated.

502.4 ***Pre-Operational Check***

Measurements of significant dimensions and parameters should be made and compared with manufacturer drawings for mixer and vessel.

502.4.1 **Dimensional**

502.4.1.1 Impeller diameter, blade width and angle, and number of blades.

502.4.1.2 Shaft length and diameter, extra keyway.

502.4.1.3 Impeller location(s) on the shaft.

502.4.1.4 Clearance between impeller and vessel bottom.

502.4.1.5 Baffle: number, width, length, and spacing from the tank wall.

502.4.1.6 Shaft runout at bottom.

502.4.1.7 Vessel dimensions: diameter, height from inside bottom to mounting face, etc.

502.4.1.8 Liquid levels.

502.4.2 **Equipment**

502.4.2.1 Motor speed, power, electrical classification, etc. and any other useful information on the nameplate.

502.4.2.2 Gear-drive reduction ratio and power rating.

502.4.2.3 Equipment near the impeller.

502.4.2.4 Equipment adjustment: bolts, belts, mounts, couplings, etc.

502.5 ***Hand Turning***

Frequently the motor coupling can be turned by hand, or with a suitable manually-operated lever, to produce slow motion of the impeller. This technique may also be used to check: direction of rotation, mixer shaft runout, obstructions in the vessel. Smooth rotation should indicate no mechanical defects in the drive equipment or seal. This test should be conducted before powered operation.

502.6 ***Jog***

Before operational start-up, the mixer may be "jogged" or "bumped," meaning rapidly turned on and off. Direction of rotation can be observed along with smooth operation without unusual noise suggesting the absence of obstructions or mechanical defects.

502.7 ***Hydrostatic Pressure Tests***

Some mixers involve pressurized lubricant reservoirs and/or pressure in the mixing vessel. Seals and containers may be given a hydrostatic or all-liquid pressure test before pressurization with a vapor. If all vapor is purged from the system before pressurization, the danger of explosive failure will be reduced since the pressure will drop rapidly if the seal or container should fail.

502.8 ***System Test in Air***

Before operating the mixer with process fluid the system may be given a mechanical running test in air or water to verify mechanical soundness and freedom from vibration. The water test may also be used to verify proper motor loading. See next Sec. 502.9 for a water test.

Caution: not all mixers are designed to operate in air.

Check with the manufacturer before running a test in an empty vessel. Possible problems include:

502.8.1 Mixer shaft may be unstable in air.

502.8.2 Variable speed clutch may require a load to operate properly.

502.8.3 Submerged steady bearing may require fluid in vessel for lubrication.

502.8.4 Submerged mechanical seals may require fluid in vessel for lubrication. This is common for side-entry mixers with single mechanical seals.

502.8.5 All mixer system components should be operated, including:

502.8.5.1 Gearbox lubrication pumps.

502.8.5.2 Coolers.

502.8.5.3 Seal lubricator.

502.8.5.4 Steady bearing lubricator.

502.8.5.5 An air running test would normally be of short duration, five minutes to 30 minutes, to do preliminary checks on: vibration, noise, runout, speed, and direction.

502.9 ***System Water Test***

Caution: not all mixers are designed to operate in water.

Check with the manufacturer before running a test in water.

502.9.1 **Possible problems include:**

502.9.1.1 Water may be an unsuitable lubricant for a submerged steady bearing or submerged mechanical seal.

502.9.1.2 The mixer may be designed to operate in a fluid that draws less power than water, and would overload the motor in water.

502.9.1.3 The materials of construction may not be corrosion-resistant to water.

502.9.1.4 Mixers designed to operate in high viscosity liquids may splash and vibrate excessively in water.

502.9.1.5 Exposure of process fluids to residual water may not be tolerated.

502.9.2 **Typical water tests include:**

502.9.2.1 All mixer system components should be operated as listed in Sec. 502.8.5.

502.9.2.2 A water batch test would usually be run for four or more hours, allowing equipment temperatures to stabilize and then do tests on the following: motor power, mixer speed, direction of rotation, runout, gear tooth contact patterns, seal leakage, auxiliary equipment, vibration, noise, and temperature, as described in the next sections of this testing procedure.

503.0 ***Operating Performance Tests***

The liquid surface location should be checked. Many mixers are not designed to operate with the surface near or just above the level of the impeller. Refer to the mixer instruction manual.

503.1 ***Speed***

Measure within 1 percent accuracy and record. Check direction of rotation.

503.2 ***Power***

503.2.1 **Start-up**

Power draw during the start-up period is often greater than during steady-state operation. Extra power is required to accelerate the fluid to establish hydraulic equilibrium. Normally this additional power is within design limits for the motor. For solids suspension, it is conceivable that the impeller could be surrounded by compacted solids, in which case start-up power could be several times greater than the steady-state operating power. Excessive power requirements may cause motor overloads and even shaft or gear failures under extreme circumstances. If an impeller is buried in solids, an air lance or other means may be needed to free the impeller.

503.2.2 **Steady State**

In long-term operations, the power draw of the mixing impeller has a mean and a variable component. The closed vessel and the unsteady nature of the flow reflected back into the impeller zone, cause a fluctuating component of impeller torque. For fast response instrumentation, the variable component of the power draw could range from 10 percent to 20 percent of the mean.

The actual magnitude would depend on installation geometry and power intensity of the mixer. These variations might not be observed for the slow response of highly damped sensors or instruments.

503.2.3 **Free Surface**

During draw-off, or operation where the impeller is near the free surface, the amount of the fluctuating compared with the mean component of power increases dramatically.

503.2.4 **Process Factors**

Mixer mean power or power variations can be affected by process related factors. Such factors might include: inlet and outlet flows, recirculation flows, equipment proximity, pressure, temperature, gas usage, liquid viscosity changes, etc.

503.2.5 **Variable Process Conditions**

Consider effects of changing liquid level, viscosity, speed, flow rates, and other process conditions. Include power failures, coagulating batches, viscosities that change with time, phase changes, and other possible variable conditions.

Changing liquid levels are an almost certainty with batch processes. Not all mixers can be operated safely while the tank is filling or emptying, check with the manufacturer. When an impeller operates near the liquid surface, splashing and random loads are likely. Batch conditions should avoid liquid levels that cause problems with impellers at the liquid level, otherwise the impeller locations must be changed.

503.3 ***Torque***

All of the factors noted in the previous section on Power apply equally to torque. Additionally, torque variations may also result in speed fluctuations, so speed measurement should be an integral part of torque measurements if the results are to be converted to power results.

504.0 *Mechanical Condition Tests*

504.1 *Alignment and Adjustments*

The procedure of measuring alignments should be obtained from the mixer or component manufacturer.

504.2 *Runout*

Typical static shaft runout at unsupported locations along the shaft should be less than 0.030 inches runout/foot (2.5 mm/m) of shaft length from the primary support. Allowable runout at seals and steady bearings depends on the type of seal, size of shaft, and operational speed. Typically, such limitations reflect maximum deflections for the peripheral speeds between rotating and non rotating parts.

504.2.1 Shaft runout may be measured while the mixer shaft is hand-turned.

504.2.2 A measurement may be made of shaft movement normal to its axis.

504.2.3 A measurement should be made at more than one radial location, such as every 90 degrees.

504.2.4 Runout can be measured at the seal and at the impeller.

504.3 *Gear Tooth Contact Patterns*

The gear drive manufacturer may provide instructions in the service manual to measure gear tooth contact patterns. The procedure may vary depending on gearing and gearbox design, but a typical procedure is as follows:

504.3.1 Drain gear box oil.

504.3.2 Remove gearbox inspection plate.

504.3.3 Wipe oil from several gear teeth.

504.3.4 Apply a thin layer of transfer dye to the gear teeth.

504.3.5 Replace inspection plate and oil.

504.3.6 Run the gear drive briefly either loaded or unloaded.

504.3.7 Drain oil, remove inspection plate, and record the pattern of dye removed, which is the gear tooth contact pattern.

504.4 *Seals*

A seal may be operated at design and/or normal pressure and temperature for a test. If seal leakage is measured over a long time, the tests may be done during normal operation and use of the mixer.

Some seals are provided with an external leakage test or drain port, which may be isolated by a low pressure secondary seal, and this can be used to collect leakage out of a vessel. Other seals are provided with an internal catchall and drain, and this can be used to collect leakage into the vessel.

A continuous log of seal lubricant replenishment may be kept during normal operation. This log can be used to calculate seal leakage.

504.5 *Auxiliary Equipment*

504.5.1 **Steady Bearings**

Before installing a steady bearing, the mixer shaft should be turned by hand at the motor coupling. The shaft end will usually scribe a circle when rotated one full turn. The steady bearing housing should be installed concentric with the center of this circle.

504.5.2 **Variable Speed Clutch**

The manufacturer's instructions should be followed to operate a variable speed clutch. Run through specified speed range and measure mixer shaft speed. Sometimes a load is required to operate a variable speed device, and tests may have to be done with water or process fluid in the mixing vessel.

504.5.3 **Lubrication and Cooling Systems**

Pressure measurements should be taken during operation according to the manufacturer's instructions.

The flow rate in forced pumping or thermosyphon systems can be difficult to measure directly.

Direction of flow and adequacy of flow can sometimes be deduced by measuring inlet and outlet temperatures of the system.

504.6 *Vibration*

If vibration measurements are being considered because of gear drive rocking, then the first step is to measure how much rocking or vibration. If the motion exceeds the manufacturer's allowable limits, more detailed tests are required.

Vibration measurements should include basic information on magnitude and frequencies. The frequency ranges should be established to highlight shaft speed, blade passage, and other identifiable frequencies.

504.7 *Noise*

Noise is a result of mechanical actions similar to vibration being transmitted to the air as pressure pulsations. Procedures for noise measuring instruments may be obtained from their manufacturers.

504.8 ***Temperature***

The temperature of gear drive oil and/or motor surface temperature should be measured after running at rated speed under load for four hours. Either water or process fluid should be in the mixing vessel.

If the gear drive is splash lubricated or internal pump lubricated, the sump oil temperature should be measured.
If the gear drive has an external pump and/or cooler, both drain and inlet oil temperatures should be measured.

Measure ambient air temperature when equipment temperature is measured.

505.0 *Mechanical Operation Tests*

Before resorting to extensive testing, be sure that the equipment has been properly installed and that there has been no damage to the equipment. A complete dimensional check and static runout measurements should be made before proceeding. Discrepancies noted in these steps often lead to a solution to vibration problems.

505.1 ***Natural Frequency***

Shaft natural frequencies are obtained with the mixer turned off.

CAUTION: for safety, the motor power supply should be locked out.

Some kind of disturbing force is required to displace the shaft and then let it oscillate at its natural frequency. Frequency measurement requires some sort of displacement transducer to detect motion and an analyzer, such as an oscilloscope to find frequency.

Natural frequency measurement should be made of the mixer support structure and components in the vessel.

505.2 ***Deflections of Structure***

These tests are done with the mixer operating in process fluid. Do not operate the unit in air or water without checking with mixer designer/manufacturer.

505.2.1 **Mixer Support Structure**

Measure deflection, twist, etc.

505.2.2 **Mixer**

Motion relative to support structure or other stationary equipment. Check for rocking at the mounting bolts.

505.2.3 **Mixer Shaft**

The mixer shaft measurements involve either strain or deflection. For deflection measurements runout must be measured and subtracted. If possible, spectral analysis of the shaft data should be obtained.

505.2.4 **Operating Modes of Vibration**

Operating system vibration magnitudes and frequencies should be measured on the tank structure and mixer housing.

506.0 *Process Condition Tests*

General test criteria and duration should be defined and frequency of data sampling established. Because of the diversity of processes and materials, this procedure cannot recommend specific sampling methods. Sampling may involve potential exposure of personnel and the environment to hazardous substances. Any sampling or field measurements must conform to plant procedures for safety and environmental control For further information about field instruments, consult other reference sources, such as the Instrument Engineers Handbook, Ref. 808.23.

All data should be identified as to the source and recorded in sufficient detail. An immediate review of the data should be carried out and the results evaluated. Then the test should be continued until the original plan is completed.

506.1 *Blending*

Blending applies primarily to the combination of miscible liquids to obtain uniform properties or uniform concentrations of ingredients. Measurements are usually applicable to batch operations, where properties become more uniform as mixing proceeds or continuous operations, where two or more streams are mixed and withdrawn at constant flow rates.

506.2 *Heat Transfer*

Heat transfer in an mixed tank is usually accomplished by exchange between the process fluid and some heat transfer fluid, contained in a jacket at the tank wall or pipe coils inside the tank. The exact methods of making heat transfer measurements depend on the test capabilities available with the installed equipment.

506.2.1 Heat rejection or addition can usually be measured by making a heat balance on the heat transfer medium. Accurate measurement of temperatures and flows into and out of the jacket or coils should provide an indication of the heat transfer rate, plus any ambient losses.

506.2.2 If the rate of heat generation by the vessel contents can be determined, a further measure of the heat transfer rate can be made, and checked against the external measurements. Alternatively, a transient study can be done to establish the rate of heat addition or removal from a known quantity of material in the tank.

506.2.3 Careful measurements of vessel contents temperatures and heat transfer fluid temperatures are necessary to convert heat transfer rates to an overall heat transfer coefficient. The coefficient may be a local or overall value depending on the temperatures or temperature averages used.

506.2.4 Heat transfer rates at actual operating conditions with real process fluids are usually the most significant, but the most difficult to obtain. Heat transfer tests with water or other inert fluids are easier and may be useful depending on the exact nature of the desired results.

506.2.5 If the mixing equipment can be operated at different speeds, different heat transfer rates would be expected. The incremental change in heat transfer will depend on the limitation due to process-side heat transfer coefficient.

506.2.6 If the process fluids are viscous, or the mixing is intense, heat addition associated with the dissipation of input mixing horsepower cannot be neglected.

506.3 ***Immiscible Liquids Dispersion***

Applications include processes where two immiscible liquids are contacted to promote mass transfer between phases or to create an emulsion. Most mass transfer operations are carried out on a continuous basis, in a series of mixer/settlers or a staged column. Emulsions may be made either batchwise, continuously or by recycling.

The purpose of a mixer is to make a liquid-liquid dispersion of a desired drop size range and at a desired production rate. Other things including process chemistry and upstream equipment affect the apparent operation of the mixer. It is very important that the test differentiates between the effect of the mixer and the rest of the process. Measurements and testing procedures may vary considerably depending on the process and operating conditions.

506.3.1 Dispersion tests measure the capability of the mixer to make drops of the correct size. The result of such a test could be a mass transfer coefficient or the number of theoretical mass transfer stages for a continuous process..

506.3.2 Hydraulic capacity tests measure the volumes of the phases that can be processed. The result could be the operating limits before entrainment of the dispersed phase becomes significant, which may occur in a staged column.

506.3.3 Emulsion tests would measure the stability of the emulsion and the droplet size distribution. These tests could be conducted before, during, and/or after a batch process or at the outlet of a continuous process.

506.3.4 Other factors including process chemistry and upstream equipment affect the apparent operation of the mixer.

506.4 *Liquid-Solid Contacting*

Applications include all processes where an impeller is used to disperse a solid into a liquid, maintain suspension, or promote mass transfer between the phases.

Caution: most mixers are not designed to start in settled solids or operate through fluid draw-off.

In these cases, test operations beyond the design limits could overload the motor or break mixer components.

506.4.1 To interpret results, characterizing the solids and the slurry is usually necessary. Measurements of particle size distribution, material density, and settling rates should be considered. The bulk slurry density is another important measurement. Slurry properties are important for mixer power evaluation since power draw increases in proportion to slurry density.

506.4.2 For continuous flow processes, the tests must monitor slurry densities of the incoming and exiting process streams. Solids accumulation or depletion could grossly change the slurry density in the vessel being tested. Such a change in density would affect mixer power and other process results.

506.4.3 Process tests could include solids suspension capability, mass transfer, material wetting characteristics, and changes in particle size.

506.5 ***Liquid-Gas Contacting***

Applications include all processes where the mixer promotes contact between a liquid and gas. These processes may include the presence of solids also. In the latter case, the section on liquid-solid contacting should also be reviewed for applicability.

Some mixers are designed with gas rate limitations. Operation at a reduced gas rate could overload the unit. For other units, operation above a prescribed gas rate could cause large fluid forces with the attendant increases in shaft deflections. The operating limit must be determined and observed.

506.5.1 The term "flooding" is often used to identify a condition where too much gas is supplied. No single definition of "flooding" exists. The definition depends on the type of impeller and the application. Impellers usually have an observable increase in torque when they become flooded. Observation of this increase in torque may be the most practical method for determining flooding in large mixers. If a process criterion uses the term “flooding,” further description is required to be sure that all parties understand what is meant. “Flooding” is not defined for up-pumping axial flow impellers.

506.5.2 To interpret results, measurement of gas rate, gas hold-up, interfacial tension, and other liquid phase and gas phase properties including density, viscosity, temperatures, pressures, etc. is usually necessary. In some processes gas evolution is used for batch temperature control. Independent procedures might be required to calculate or measure the rate of evolution.

506.5.3 Gas rate can have a significant impact on mixer power. It is sometimes desirable to measure ungassed and gassed power. This ratio is often called the gassed power factor.

506.5.4 Process tests could include gas dispersion capabilities, flooding, gas hold-up, mass transfer, etc. Mass transfer testing techniques have been developed for steady state and unsteady state processes. If processes cannot be run under process conditions, air-water tests might be applicable.

506.6 ***Liquid-Gas-Solid Tests***

Solids suspensions, as low viscosity suspensions or slurries, should have a small effect on gas dispersion. Therefore, typical gas dispersion tests for flooding and bubble size should be possible even with solids present.

The presence of gas will usually adversely affect the ability of a mixer to suspend solids. A suspension level test should be conducted at the maximum gas rate.

Although multiphase tests are extremely complicated, no general procedures specific to this situation are offered. Most tests for simpler systems may be applicable, and some combination of these tests may be appropriate for a given condition.

506.7 ***Other Tests***

If tests in the full scale process equipment are impractical, or do not reveal the cause of problems or possible methods for solution, small scale modeling tests may be the next best approach. Although small scale modeling, scale-up, and scale-down are beyond the scope of this procedure, the methods must be considered as an alternative.

Often, pilot or laboratory scale mixing equipment may be available. Tests done in transparent tanks with variable speed mixers can provide enormous insight into many mixing and agitation problems with simple equipment.

Many tests described in this procedure can be applied to pilot plant equipment with greater ease, less cost, and better control. Changes to the equipment can be made in less time and at less cost than with larger equipment.

Properly interpreted and scaled-up, the results of small scale testing can be of enormous benefit. Any test program for impeller mixing equipment should consider small scale testing as a possible alternative to large process tests.

600.0 Computation of Results

601.0 *Data Requirements*

All relevant dimensions, operating conditions and results should be recorded, showing the appropriate units of measure. If the accuracy or measurement technique associated with a result might be questioned later, a notation should be made. Readings that fluctuate during a test may be recorded as an average, possibly along with a minimum and maximum. If the variation is significant or an accurate average is required, some sort of monitoring and averaging techniques may be considered.

601.1 ***Dimensional Measurements***

All dimensional measurements (Sec. 502.3) should be verified and recorded. Simple differences between actual and design impeller diameter can account for major process and mechanical failures.

601.2 ***Operating Conditions***

Pressure temperature, liquid level(s), feed and discharge rates, and output shaft speed should be recorded, along with any special variable measurements.

601.3 ***Process Properties***

Fluid properties, such as density and viscosity should always be determined and recorded. Other properties such as solids size and density, gas flow rates, and thermal properties may also be determined depending on the type of test being done.

602.0 *Fundamental Calculations*

Certain fundamental calculations are recommended to help evaluate operational performance for most mixing systems. The following list of relationships is not comprehensive, but should suffice for many situations. Other relationships with more specific meaning are presented where appropriate.

602.1 ***Reynolds Number***

$$Re = \frac{D^2 N \rho}{\mu}$$

D	=	impeller diameter
N	=	rotational speed
ρ	=	fluid density
μ	=	viscosity

602.1.1 English units

$$Re = \frac{10.74\, D^2 N (S.G.)}{\mu}$$

D	[inches]
N	[rev/min]
$S.G.$	specific gravity
μ	[centipoise]

602.1.2 SI metric units

$$Re = \frac{D^2 N \rho}{\mu}$$

D	[m]
N	[rev/s]
ρ	[kg/m^3]
μ	[Pa s]

602.1.3 typical values:

Turbulent Mixing:	$Re > 20{,}000$
Transition:	$10 < Re < 20{,}000$
Laminar Mixing:	$Re < 10$

602.2 *Power Number*

$$Po = \frac{P\, g_c}{\rho N^3 D^5}$$

P	=	impeller power
ρ	=	density
N	=	rotational speed
D	=	impeller diameter

602.2.1 English units

$$Po = \frac{1.524 \times 10^{13}\, P}{(S.G.)\, N^3 D^5}$$

P	[horsepower]
$S.G.$	specific gravity
N	[rev/min]
D	[inches]

602.2.2 SI metric units

$$Po = \frac{P}{\rho N^3 D^5}$$

P	[W]
ρ	[kg/m^3]
N	[rev/s]
D	[m]

602.2.3 typical values (depend on impeller type):

Low: $0.1 < Po < 0.5$
for axial flow impellers: propellers, hydrofoil impellers, etc.

Middle: $0.5 < Po < 3.0$
for mixed flow impellers: pitched-blade turbines, simple paddles, etc.

High: $3.0 < Po < 7.0$
for radial flow impellers: straight-blade turbines, disc-style turbines, etc.

602.3 ***Torque***

$$\tau = \frac{P}{2\pi N}$$

τ = torque
P = power
N = rotational speed

602.3.1 English units

$$\tau = \frac{63{,}025\,P}{N}$$

τ [inch-pounds]
P [horsepower]
N [rev/min]

602.3.2 SI metric units

$$\tau = \frac{P}{2\pi N}$$

τ [N·m]
P [W]
N [rev/s]

602.4 ***Mixer Speed***

$$N = \frac{N_R}{t}$$

N_R = number of shaft revolutions
t = time interval in which shaft revolutions are counted

602.4.1 English units

$$N = \frac{N_R}{t}$$

N [rev/min]
N_R [dimensionless]
t [min]

602.4.2 SI metric units

$$N = \frac{N_R}{t}$$

N [rev/s]
N_R [dimensionless]
t [s]

603.0 ***Operating Performance Calculations***

Perhaps the simplest performance evaluation is motor loading. Electrical measurements can be compared directly with the manufacturer's performance curves. A typical motor performance curve is shown in Fig. 805.4, and a sample calculation is shown in Sec. 805.1.5. Similarly, actual performance data can be obtained with a watt meter and interpreted.

603.1 *Power*

Simple calculations convert voltage and amperage to input electrical power. This information, in combination with motor performance curves or experimentally measured power factors can determine a motor loading, or the fraction of available motor power being delivered to the mechanical equipment.

Mixer power can be related to power and/or torque plus speed measurements by factoring the efficiencies of all the interceding components. Typical components include: electric motor, gear drive, couplings, belts and pulleys, torque limiters/clutches, shaft seals, bearings and shaft supporting devices, etc.

Since power is a magnitude parameter, strongly related to the size of the equipment involved, power compared with the amount of process fluid is frequently more relevant.

603.1.1 **Power per Unit Volume**

$$\frac{P}{V} = \frac{P_m L_f E_m}{V}$$

P_m = motor power rating
L_f = motor loading, fraction
E_m = mechanical efficiency of couplings, gearbox, and seal, fraction
V = fluid volume

603.1.1.1 English units

$$\frac{P}{V} = \frac{1{,}000\, P_m L_f E_m}{V}$$

P/V [horsepower/1,000 gallons]
P_m [horsepower]
L_f [fraction]
E_m [fraction]
V [gallons]

Typical magnitudes in English units of hp/1,000 gal are shown below.

Low:	0.1	<	P/V	<	0.5
Middle:	0.5	<	P/V	<	3.0
High:	3.0	<	P/V	<	15.0

Low values typically are used in large storage tanks with low viscosity liquids. High values are found in chemical reactors, where mixing improves productivity.

603.1.1.2 SI metric units

$$\frac{P}{V} = \frac{P_m L_f E_m}{V}$$

P/V	[W/m^3]
P_m	[W]
L_f	[fraction]
E_m	[fraction]
V	[m^3]

Typical magnitudes in metric units of W/m^3 are shown below.

Low:	20	<	P/V	<	100
Middle:	100	<	P/V	<	600
High:	600	<	P/V	<	3,000

The mixer speed can be measured directly or calculated based on motor speed and drive reduction, provided the exact gear ratio is known or measured. Standard motor curves give slip versus motor load, to predict actual motor speed. The standard gear drive speed ratio values based on the AGMA rating procedure will be within 2-4 percent of the true ratio. Some mixer manufacturers use exact ratios, or the ratio may be determined by counting input and output shaft revolutions, or knowing numbers of gear teeth.

A complete analysis of the mixer power draw is more complex. The equipment designer should have historic data to predict the impeller power draw for standard conditions. Many references (including Ref.

808.10.2, 808.9, 808.12, and 808.7.3) found in the literature have power draw data for mixing impellers. Nonstandard process factors and nonstandard installation geometry factors must also be considered.

Typical geometry factors likely to influence power requirements include: impeller proximity to surface, bottom, side walls or any nearby obstruction, unusual tank shape or baffle arrangement etc. Geometric variables should always be accurately measured for future evaluation.

603.2 ***Torque***

From mixer speed and impeller power the torque applied to the process can be determined. For most blending applications, torque per volume is considered a better comparative measure of mixing intensity than the corresponding power per volume information.

603.2.1 **Torque per Unit Volume**

$$\frac{\tau}{V} = \frac{P_m L_f E_m}{N V}$$

P_m = motor power rating
L_f = motor loading [fraction]
E_m = mechanical efficiency of couplings, gearbox, and seal, [fraction]
N = rotational speed
V = fluid volume

603.2.1.1 English units

$$\frac{\tau}{V} = \frac{63{,}025\, P_m L_f E_m}{N V}$$

τ/V [inch-pounds/gallon]
P_m [horsepower]
L_f [fraction]
E_m [fraction]
N [rev/min]
V [gallons]

Typical magnitudes in English units - inch-pounds/gallon, low values for low viscosity fluids and high values for high viscosity fluids are shown below.

Low:	0.05	$<$	τ/V	$<$	0.2
Middle:	0.2	$<$	τ/V	$<$	1.5
High:	1.5	$<$	τ/V	$<$	10

603.2.1.2 SI metric units

$$\frac{\tau}{V} = \frac{P_m L_f E_m}{2\pi N V}$$

τ/V	[N m/m^3]
P_m	[W]
L_f	[fraction]
E_m	[fraction]
N	[rev/s]
V	[m^3]

Typical magnitudes in metric units -Nm/m^3, low values for low viscosity fluids and high values for high viscosity fluids are shown below.

Low:	1.7	$<$	τ/V	$<$	6.9
Middle:	6.9	$<$	τ/V	$<$	5.1
High:	5.1	$<$	τ/V	$<$	343

603.2.2 **Torque per Specific Volume**

$$\frac{\tau}{V^*} = \frac{(\tau/V)}{S.G.}$$

V^* = specific volume
$S.G.$ = specific gravity

604.0 *Mechanical Condition Calculations*

Most alignment, adjustment and vibration measurements are comparative. Specific calculations may be required to interpret certain results, but general calculation procedures do not exist.

605.0 *Mechanical Operation Calculations*

605.1 *Natural Frequency*

The natural frequency of an impeller shaft depends on shaft length, impeller weight, bearing spacing, and material properties.

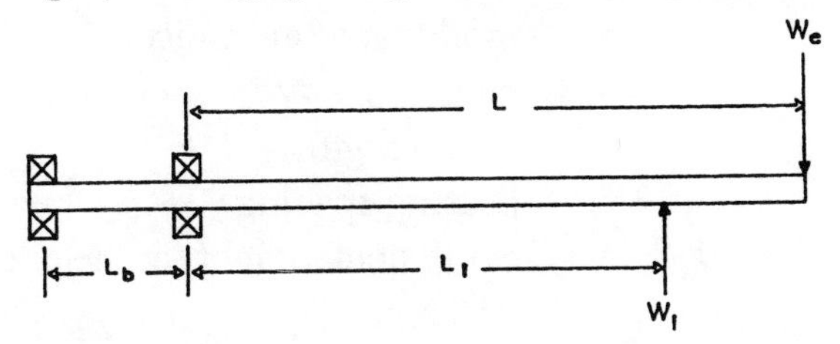

Fig. 605.1 Shaft for Natural Frequency

The equation for natural frequency (critical speed) of a constant-diameter, overhung beam (like shown in Fig. 605.1) can be written as follows for English units:

605.1.1 **Natural Frequency (Critical Speed, N_C)**

$$N_C = \frac{d^2 \sqrt{E_Y/\rho_m}}{L\sqrt{W_e}\sqrt{L+L_b}}$$

$$= \frac{d^2}{L}\sqrt{\frac{E_Y/\rho_m}{W_e\left(L+L_b\right)}}$$

d = shaft diameter
E_Y = modulus of elasticity
ρ_m = metal density
L = shaft length
L_b = bearing spacing
W_e = equivalent weight (at the end of shaft of length, L)

$$W_e = \sum_i W_i\left(L_i / L\right)^3 + w_s L/4$$

L_i = impeller location (down shaft)
W_i = impeller weight
w_s = shaft unit weight

605.1.1.1 English units

$$N_C = 37.8 \frac{d^2 \sqrt{E_Y/\rho_m}}{L\sqrt{W_e}\sqrt{L+L_b}}$$

$$= 37.8 \frac{d^2}{L}\sqrt{\frac{E_Y/\rho_m}{W_e\left(L+L_b\right)}}$$

N_c [rev/min]
d [inches]
E_Y [psi]
ρ_m [lb_m/in^3]
L [inches]
L_b [inches]
W_e [lb_m]

$$W_e = \sum_i W_i\left(L_i / L\right)^3 + w_s L/4$$

L_i [inches]
W_i [lb_m]
w_s [lb_m/in]

605.1.1.2 SI metric units

$$N_C = 0.134 \frac{d^2 \sqrt{E_Y/\rho_m}}{L\sqrt{W_e}\sqrt{L+L_b}}$$

$$= 0.134 \frac{d^2}{L} \sqrt{\frac{E_Y/\rho_m}{W_e\left(L+L_b\right)}}$$

N_C [rev/s]
d [m]
E_Y [Pa]
ρ_m [kg/m^3]
L [m]
L_b [m]
W_e [kg]

$$W_e = \sum_i W_i\left(L_i / L\right)^3 + w_s L/4$$

L_i [m]
W_i [kg]
w_s [kg/m]

Conventional calculations reflect only static analysis of the shaft. Such calculations are adequate for most mixing equipment. Equipment with large, high speed shafts may require dynamic analysis.

For more complex shafts or ones with more than two bearing supports, contact the mixer designer/manufacturer.

605.2 ***Deflection***

Two principal forces contribute to shaft deflection. Fluid forces are the result of unsteady hydraulic flow in the region of the impeller. The second is centrifugal force due to the combined effect of shaft runout and mechanical imbalance.

Fluid force data are not available for all types of impeller. Each impeller has a characteristic force number (in similar fashion to power and flow numbers). The mixer designer/manufacturer should be contacted for standard fluid force numbers or design hydraulic loads. An estimate of hydraulic force can be obtained from the following formula.

605.2.1 Lateral Hydraulic Force

$$F_f = \frac{P_m f_s}{N D}$$

F_f = random lateral hydraulic force
P_m = motor power
f_s = hydraulic factor
N = shaft speed
D = impeller diameter

605.2.1.1 English units

$$F_f = 19000 \frac{P_m f_s}{N D}$$

F_f [lb_f]
P_m [hp]
f_s dimensionless
N [rpm]
D [inches]

605.2.1.2 SI metric units

$$F_f = 0.048 \frac{P_m f_s}{N D}$$

F_f [N]
P_m [W]
f_s dimensionless
N [rpm]
D [inches]

Motor power is recommended because process conditions may result in full load, even if impeller design is less. Power should be split between multiple impellers in proportion to their relative power draws. The hydraulic factor, f_s, may be 1.0 for simple blending with some open style impellers. This factor may be 3.0 or greater for conditions, such as gas dispersion, that create higher random loads. Process (503.2.4) and geometry (603.0) factors for nonstandard installations also affect the hydraulic factor.

The centrifugal forces can be readily calculated by measuring runout and estimating mechanical imbalance.

A simple beam analysis will give shaft deflection versus forces applied at the impellers(s).

606.0 *Process Condition Calculations*

606.1 *Blending*

A first step in determining blending results is complete documentation of the test conditions, the measurement locations, and the sensing or sampling equipment used. Observation of blend time may be highly dependent on minor factors, like the measurement location or method of initiating the blend test. Times may be unnecessarily extended by poorly mixed locations on the liquid surface or in exit nozzles.

True uniformity is difficult to measure since it is dependent on both time and location. However, use of a pH or an ion electrode may work to detect the final concentration resulting from a pulse addition. Other sampling techniques may work in large vessels, where blend times are long.

Typical blend time results might be obtained by measuring the time required for uniformity in five to ten repeated tests. Variability between runs is typically at least 10 percent because of the random nature of mixing. An average time may be combined with the impeller rotational speed to establish a dimensionless number, especially if scale-up is involved.

606.1.1 Dimensionless Blend Time

$$\Theta = t_{blend} N$$

t_{blend} = blend time
N = rotational speed

Caution must be exercised in the use of a dimensionless quantity obtained by this means, since it assumes that at least geometric similarity is retained in any other situation. Important factors, such as the impeller-to-tank-diameter ratio, must be the same for other conditions. Unless process conditions are fully turbulent, the effects of Reynolds number must also be considered.

606.2 *Heat Transfer*

Specify all properties, geometry, and operating conditions and use the results in computations as needed. The essential properties are specific heat, specific gravity, thermal conductivity and viscosity at specified temperatures and compositions. Additional tests may be needed if actual conditions differed from design conditions.

606.3 *Immiscible Liquid Dispersion*

Liquid dispersion mechanisms are complex, since both mixing intensity and fluid properties influence results. Fluid properties that influence process results include density, viscosity, and interfacial tension. The dispersion of two liquids is also influenced by which phase is continuous and which is dispersed.

In spite of the fluid complexities, the tip speed of a given impeller seems to affect the final drop size. Although no simple rules exist for drop size, determination of impeller tip speed is probably a good operational variable to record in conjunction with dispersion.

606.3.1 Impeller Tip Speed

$$v_{ts} = \pi N D$$

N = rotational speed
D = impeller diameter

606.3.1.1 English units

$$v_{ts} = 0.2618\,N\,D$$

v_{ts} [ft/min]
N [rev/min]
D [inches]

typical magnitudes (English units - ft/min)

Low:	200	$< v_{ts} <$	500
Middle:	500	$< v_{ts} <$	900
High:	900	$< v_{ts} <$	2000

606.3.1.2 SI metric units

$$v_{ts} = 0.5\,N\,D$$

v_{ts} [m/s]
N [rad/s]
D [m]

typical magnitudes (metric units - m/s)

Low:	1	$< v_{ts} <$	2.5
Middle:	2.5	$< v_{ts} <$	4.5
High:	4.5	$< v_{ts} <$	10

Material balances are also important in liquid dispersion. To determine the equipment performance and the adequacy of procedures, select data from known periods of steady operation. Steady state can be determined from charts of flow rates, valve positions, and decanter levels. The balances should be calculated on as many stream species as possible. The balances can be checked by comparing predicted readings with the actual measurements.

Entrainment of one phase into the other is a common cause and/or result of poor hydraulic performance in continuous systems. When the volume or flow rate of the continuous phase overwhelms the mixing, the dispersed phase material may be carried through the mixer The important data include the flow rates and the measurement of one phase contained in the discharge stream of the other phase.

606.4 *Liquid-Solid Contacting*

Material balances are important to determine the equipment performance. The balances can be used to determine the approach to steady state and to see if significant particle classification is occurring. Select the data from known periods of steady inflow and outflow. These periods can be determined from flowmeter readings, level readings, and valve positions. The balances should be calculated on both the overall quantity of material and on the individual particle size ranges.

Mixer capacity should be measured by means of material balances. The data should be selected from periods of steady operation. This means not only steady inflows and outflows but also no accumulation. The capacity can be affected by upstream process conditions. The process materials should be that used for the actual operations.

606.5 *Liquid-Gas Contacting*

The gas hold-up should be measured during steady state operation. Select data from steady periods of operation. The hold-up is calculated by the difference in apparent liquid volume between ungassed and gassed periods of operation. Mass transfer measurements can be made during steady operation or by a transient technique. For the steady state operation tests, select data from periods of well-established steady operation. For transient tests, duplicate runs should be selected as a check for the reproducibility of the data.

606.6 *Other Tests*

Other calculations may be required for other tests, such as interpretation of concentration results as rates of reaction.

700.0 Interpretation of Results

701.0 *Introduction*

The test results should be interpreted and evaluated for the specific objective of the test and the accuracy required.

702.0 *Interpretation of Operating Performance*

702.1 *Speed*

The measured speed should be compared with the predicted value. The difference should be within the error of measurement. Impeller power is strongly dependent upon speed. The table below shows how the errors in predicting speed affect the prediction for power.

	Power Error	
Speed Error	Laminar ($Re < 10$)	Turbulent ($Re > 20.000$)
5%	10%	16%
10%	21%	33%

702.2 *Power*

The measured power should be within approximately 8 percent of the calculated value for normal conditions.

Greater variations are usually the result of complicating process factors or installation geometry factors. Sometimes a process upset results in much higher mixer loads. The unit design is generally based on the worst case condition. Under normal conditions the mixer power draw may be much lower than the capability of the motor and drive.

702.3 *Torque*

The product of torque and speed with the appropriate conversion factor (See Sec. 602.3) should be within 10 percent of the predicted power. Greater variations are usually the result of complicating process factors or installation geometry factors.

Sometimes a process upset results in much higher mixer loads. The unit design is generally based on the worst case condition. Under normal conditions the mixer power draw may be much lower than the capability of the motor and drive.

703.0 *Interpretation of Mechanical Conditions*

703.1 *Alignment and Adjustments*

Adjustments should be made until all mechanical equipment is mounted and aligned to within specifications of the manufacturer. Any deviations that cannot be corrected should be reported to the equipment manufacturer and resolved.

703.2 *Runout*

Runout measurements should be compared with manufacturer or purchase specifications. Any deviations that cannot be corrected should be reported to the equipment manufacturer and resolved.

703.2.1 Total indicated runout at a seal is commonly limited to five to 30 mils (125 to 750 microns), depending on seal type and operating speed.

703.2.2 Total indicated runout at the end of a shaft is commonly limited to 30 mils per foot of shaft (2.5 mm per meter of shaft length).

703.3 *Gear Tooth Contact Patterns*

The gear tooth contact pattern may be compared with the gear drive standards of the manufacturer. If percent contact is excessively low, the gear drive may require adjustment or replacement of parts as needed.

703.3.1 The percent contact rarely exceeds 80-90 percent under the best conditions. A gear drive may be designed to operate at lower than 80 percent contact.

703.3.2 For an interpretation of gear tooth contact patterns see: AGMA 390.03, Section 9, pages 96-99.

703.4 *Seals*

703.4.1 **Separately Mounted Seals**

Compare concentricity and squareness of the seal to the manufacturer's standards.

703.4.2 **Packing Seals**

Compare leakage rates to specifications.

703.4.3 **Mechanical Seals**

Compare leakage rates and temperature with specifications.

Seal designs differ, but the leakage rate of lubricant can be expected to be about one drop per minute per inch of shaft diameter per 100 psi pressure differential (41 ml per day per 100 mm of shaft diameter per 100 kPa pressure differential).

703.5 ***Auxiliary Equipment***

Measurements should be compared with specifications.

703.6 ***Vibration***

703.6.1 Displacement measurements may be compared with AGMA 300.01, Figure l, type A.

703.6.2 Velocity measurements may be compared with AGMA 300.01, Figure 2, type A.

703.6.3 Acceleration measurements may be compared with AGMA 300.01, Figure 3, type A.

All these measurements must be made on a very rigid base and may not be practical for most installed equipment.

703.7 ***Noise***

703.7.1 Sound pressure level measurements can be compared with motor and gearbox manufacturers' specifications.

703.7.2 Most of the noise associated with a mixer usually comes from the motor. Mixers should be designed and tested to meet OSHA standards.

703.7.3 Measurements can be compared with the OSHA and Walsh-Healey Public Contracts Act, 50-204.10. Sound level is currently limited to 90 dBA for eight hours exposure per day without protection. Lower sound limits, such as 85 dBA, may be set by individual companies.

703.7.4 Octave band measurements of excessive or unusual noise can be interpreted as to possible causes. Noise can be reduced either by dealing directly with the causes or by shielding the mixer drive.

703.8 ***Temperature***

703.8.1 Gear drive oil temperature and motor surface temperature should be compared with specifications of the manufacturer.

703.8.2 Gear drive, thermal ratings are based on a maximum sump temperature rise of 100°F (55 C) above ambient, and sump temperature not to exceed 200°F (93 C) per AGMA 420.04, Section 6.2.

703.9 ***Water Power and Corrections***

Power measurements taken with water in the vessel can be translated to an expected process power draw by making a correction for process fluid specific gravity provided the impeller Reynolds number (See Sec. 602.1) indicates turbulent conditions. For viscosities resulting in transitional or laminar conditions, a complete power number curve is necessary to estimate actual power requirements.

704.0 *Interpretation of Mechanical Operation*

The support structure should be rigid. Significant motion of this structure could be due to insufficient design rigidity for the mixer loads or dynamic interaction due to the frequency of the loads. If significant motion is observed, the structure must be reinforced and the mixer rechecked for shaft deflection.

Relative motion between the mixer and support structure could be the result of insufficient bolt-down forces.

If the support structure is rigid and no relative motion exists between the drive structure and support, measure the shaft deflections. Compare these measurements with the manufacturers design values. If shaft deflections exceed the values, several possible causes may exist:

704.1 *Natural Frequency Vibrations*

The calculated value for shaft natural frequency should be within 5-10 percent of the measured value. A difference greater than this will require a determination of the cause.

The measured natural frequency should be a sharp peaked response on the signal versus frequency plot. A broad band frequency response probably indicates improper mounting hardware tension or flexible support structure.

A direct comparison of measured system natural frequency to operating speeds is the most direct way of determining structural resonance problems. If the ratio of operating vibration frequency to system natural frequency (harmonic ratio) is between 0.8 and 1.20, then structural resonance problems are likely. Good design practice requires that this ratio should be less than 0.8. Lower operating speeds may be advisable in cases where additional loads, such as those caused by impeller operation at the liquid level, are likely to be encountered.

For large equipment, over-critical operations (ratio greater than 1.0) can create unstable speed ranges and should be used only after detailed analysis and testing. Many small or portable mixers operate above first critical speed.

Blade passage frequency can be a factor, especially for geometries involving asymmetries in the flow field. The following table shows the harmonic ratio of operating speed to natural frequency that may cause problems for impellers with different numbers of blades.

No. of Blades	Harmonic Ratio
2	0.50
3	0.33
4	0.25

Blade passage frequencies may also appear as interactions between blades and baffles.

The principle operating harmonics are shaft speed and blade passage frequency. Other harmonics could be generated by baffle interactions, sparge flows, gas, etc. The solution scheme will depend on the magnitude of the harmonic ratio. For harmonic ratios between 0.8 and 1.0 the solution would likely include stiffening the structure or lowering the operating speed. For harmonic ratios between 1.0 and 1.2, the solution above could make the problem worse. It is essential that the changes reduce the harmonic ratio to a value less than 0.8.

704.2 ***Deflection***

Compare the calculated shaft deflection with the measured values. If the sums of fluid forces and centrifugal forces (runout/imbalance) give a calculated deflection within 30 percent of the observed value, then these forces are the likely primary cause. If large forces are the cause, then the solution is to either increase the shaft diameter or reduce the forces.

Fluid forces are proportional to shaft speed squared and impeller diameter to the fourth power. Changing either would significantly alter the fluid forces.

Additionally, if geometry or process factors are causing abnormally large forces, these factors may be changed to reduce the forces.

705.0 ***Interpretation of Process Conditions***

705.1 ***Blending***

For batch mixing, a summary of properties versus mixing time can be used to see when all points lie within the range of desired uniformity. For continuous mixing, a summary of properties for each sample taken can be used to see if all points are within a desired range. Both average values and standard deviations of values can be used to evaluate batch and continuous blending test results, and may be used to estimate the frequency of data required for the remainder of a test. Interpretation of results should begin at the start of a test and continue throughout the test.

Although blend time results indicate that the time required for uniformity is inversely proportional to rotational speed for a given tank, a speed change is rarely a practical solution for improved blending. Typical increments in speed for fixed ratio gear reducers are about 20 percent. A 20 percent increase in speed results in more than a 70 percent increase in required power. Adjusting the process to the available blend time is often the only practical approach, although batch size and method of ingredient addition may help control the effects of blending.

705.2 ***Heat Transfer***

Log-mean temperature differences, actual heat transfer area and total heat transferred are essential to calculate an overall coefficient for coils and jackets. Compare the calculated overall coefficient with the design or operating coefficient. Fouling may cloud results and efforts should be made to avoid fouling by proper cleaning. Corrections for different operating conditions, heat losses and heat gains should be made. Correlations related to the Nusselt number are usually available in references on mixing.

705.3 ***Immiscible-Liquid Contacting***

The capacity of a system is the combination of the hydraulic capacity and the mass transfer rates.

705.3.1 The hydraulic capacity is usually determined by either entrainment or flooding.

705.3.2 The mass transfer capacity is determined by the residence time, interfacial area, and interfacial chemistry. The mixer design primarily affects the residence time and interfacial area.

705.3.3 Corrections to other conditions must be done with caution since the new conditions may change the interfacial chemistry. If the chemistry will not be affected, the adjustments based on volume handling capabilities, power input, and number of theoretical stages are adequate.

705.3.4 The sources of error for both types of testing include flowmeter readings, sampling error, and chemical analysis error. All these will give material balance errors.

705.3.5 The results of these tests are only valid for the range of the operating conditions studied. If changes in performance occur, then both the current operating conditions and the system chemistry must be examined to find the cause.

705.3.6 Liquid-liquid contacting results are very hard to generalize. Any test of process equipment should include consultation with the equipment vendor, the contractor designer, and process chemistry experts.

705.3.7 Frequent causes of poor performance include poor dispersion, poor mass transfer rates, and poor phase separation.

705.3.7.1 Poor dispersion is related to inadequate hydraulic performance. Contactor flooding and low power draw are indicators of this problem.

705.3.7.2 Poor mass transfer and poor separation indicate a change in the interfacial chemistry of the system. Changes in feed stream impurity concentration indicate potential problems in this area.

705.4 ***Liquid-Solid Contacting***

The capacity of a system is determined by the ability of the mixer to achieve the desired degree of solids uniformity. Once this is achieved, system chemistry becomes the limiting factor.

705.4.1 The solid suspension results can usually be extrapolated to other conditions using equations, correlations or scale-up.

705.4.2 The mass transfer results can usually be extrapolated to other conditions provided the state of the suspension will remain constant.

705.4.3 A conservative estimate of the preferred new operating conditions can be made based on equal power per unit volume.

705.4.4 The sources of error for both solid suspension and mass transfer are related to material balance error. A likely source of error is the sampling method.

705.4.5 Liquid-solid results are hard to generalize and extrapolate from one chemical system to another. Any attempt to extrapolate the results of these tests should involve consultation with the equipment vendor, designer, and recognized experts in the field.

705.4.6 A frequent cause of poor performance is poor off-bottom suspension of solids.

705.4.6.1 Possible methods to correct off-bottom suspension problems include more intense mixing (higher speed, larger impeller), reduced solids concentration, and/or smaller particle size.

705.4.6.2 Once the performance is adequate, further change in any of the variables listed above will usually give a minor improvement to the performance.

705.4.7 Another common source of problems with particulate solids, is the wetting of dry materials.

705.4.7.1 Dry materials dumped on the surface of the liquid in a mixed tank may not wet well enough to be drawn into the batch.

705.4.7.2 Changes in methods of adding materials may correct the problem.

705.4.7.3 Partial removal of baffles (at the top of the tank) or increased mixing intensity may be necessary.

705.5 ***Liquid-Gas Contacting***

The power results are usually reported as the ratio of the gassed to the ungassed power. The gas hold-up is reported as a function of the shaft power and the physical properties of the system. The mass transfer coefficient is usually reported as the combined coefficient-area term. The coefficient-area term varies with the power and the physical properties of the system. Results should be extrapolated to other conditions with caution.

705.5.1 The common sources of error are in the power measurements, in the samples for mass transfer measurements, and in knowing exactly what the level instrument reading means.

705.5.2 A common reason for poor operation is poor gas dispersion. This is normally corrected by increasing the power input to the system by means of either higher shaft speed or a larger impeller.

706.0 *Other Tests*

Interpretation of other tests must be based on the purpose of the test. Reviewing such results with appropriate experts in the field is important.

707.0 *Sources of Error*

The most common sources of errors in mixer tests are in the differences between the actual and the desired measurements. For instance, in process tests knowing how much power is being input into the fluid is important. Most methods for power measurement require some estimates for losses between the point of measurement and the process fluid. Similarly, if knowledge about the uniformity of batch composition is desired, most available measurement techniques are indirect, such as monitoring exit composition.

Additional problems arise when calculations must be done after the test data are collected or samples must be analyzed in the laboratory. A torque measurement, which will fluctuate, must be computationally combined with the operating speed to determine power. Selection of the appropriate average torque reading to go with the correct speed may not be a simple problem. Similarly, laboratory measurements of viscosity may be significantly in error compared with process conditions for several reasons, including changes with time, temperature and shear rate.

All of the other common sources of experimental error can pose problems. Instrument errors, calibration problems, recording errors, incorrect calculations or units conversion, and so forth.

708.0 *Computer Modeling*

Some computer modeling techniques are now available, the most common of which is generally termed computational fluid dynamics (CFD). Computational fluid dynamics uses either a finite element or finite difference method to model fluid motion. The mixed tank is divided into small elements in which the basic equations for continuity, momentum, etc. are solved and then connected to surrounding elements. Iterative calculations converge the model to a consistent solution for all of the fluid elements.

Several computer codes have model development packages specifically designed to create two- or three-dimensional models of mixed tanks. A variety of impeller types and tank geometries can be modeled. Although any computer model should be validated against known conditions or test results, several computer software companies have displayed the ability to adequately model flow patterns for typical mixer applications.

The CFD models may show anticipated flow patterns within a tank. Such models may show regions of poor circulation or other flow problems. The models also can be used to test how changes, such as impeller location, may affect the mixing. While modeling is not a substitute for testing, modeling may help to define a mixing problem to guide a test program.

800.0 Appendix

801.0 *Trouble Shooting Checklist*

Problems and Possible Causes

1. Unit does not turn
    - Not wired correctly
    - Mechanical blockage

2. Motor overloads
    - Mechanical blockage
    - Alignment problems
    - Motor is faulty
    - Motor undersized
    - Impeller diameter or speed wrong
    - Bearing problems
    - Viscosity and/or density too high
    - Impeller imbedded in settled solids

3. Mixer turns, but
    a. No apparent motion in the tank
        - Shaft or impeller not attached
        - Very low mixing intensity
        - Mixer incorrect for process
    b. Wrong flow patterns in the tank
        - Motor turning in wrong direction
        - Baffles incorrect or missing
        - Impellers in wrong location
        - Incorrect impellers installed
        - Mixer installed incorrectly
    c. Mixer overheats
        - Improper lubrication
        - Undersized gearbox
        - Drive belts slipping
        - Bad bearings
        - Seal problems
    d. Excessive noise in the gearbox
        - Poor lubrication
        - Bad bearings

- Loose component in the gearbox
- Gear defective or worn

4. Noise in the tank
   - Loose components
   - Debris in the tank
   - Intense turbulence
   - Cavitation

5. Excessive vibration
   - Improper support or wall thickness
   - Misalignment
   - Improper mechanical design
   - Cavitation
   - Imbalance (e.g., lost impeller blade, poor construction)

802.0 *Glossary*

For definitions of the basic equipment variables, see Sec. 203.1. Mixing related groups are shown in Sec. 203.2.

802.1 **accuracy** - The amount by which a measurement differs from the true value.

802.2 **agitation** - The random and fluctuating fluid motion resulting from the operation of an impeller-type mixer. The purpose of mixing can be liquid blending, heat transfer, solids suspension, gas dispersion, or numerous batch or continuous fluid operations.

802.3 **mixer drive** - Usually a specially designed gear reducer or other drive arrangement to reduce speed and increase torque, with heavy duty bearings on the output shaft to handle large, overhung loads.

802.4 **baffles** - Vertical plates attached to tank walls to prevent uncontrolled swirling of liquid contents.

802.5 **bearings** - A low-friction, contact point for supporting a rotating shaft in a stationary housing. Most commonly used in mixing equipment are tapered roller bearings, ball bearings or thrust bearings.

802.6 **blending** - The action of combining two materials, usually fluids in impeller-type mixing equipment, to make one uniform mixture. The time required to accomplish this action may be called **blend time**.

802.7 **cavitation** - Cavitation occurs when the local pressure, such as behind an impeller blade, is reduced below the vapor pressure of the liquid. At that pressure, a vapor or gas bubble will form, altering the flow, and often leading to the collapse of the bubble.

802.8 **coils** - To add heat transfer surface inside a mixing vessel, pipe coils can be bent in a helical spiral around the mixer. One, two, or three banks of coils may be used, however a large number of coils reduce the effectiveness of the mixer. Banks of coils may also be mounted vertically as a substitute for plate baffles.

802.9 **coupling** - A connecting device used to join two rotating shafts in a mechanical system. The purpose of a coupling is usually to provide some flexibility in alignment, but may also cushion vibration or reduce start-up torque.

802.10 **critical speed** - Rotational speed corresponding to first lateral natural frequency of a mixer shaft and impellers. (See natural frequency.)

802.11 **deflection** - The temporary bending, below the elastic limit, of a shaft, for instance.

802.12 **dimensionless groups** - A combination of process variables, normally as a ratio, in which the units of measure cancel one another. The resulting value is independent of the units used, and may have some physical significance independent of the absolute size of the equipment.

802.13 **dip pipe** - A pipe, normally attached to one of the nozzles on the top of the tank, through which material can be added or withdrawn. The dip pipe should extend below the liquid surface, during normal operation, to a region near the impeller, where the mixing is most intense. Intense mixing should provide rapid dispersion of additions and uniformity of withdrawals.

802.14 **dispersion** - The action or result of combining two immiscible fluids. In impeller mixing and agitation, dispersions may be formed with two liquids or a liquid and a gas (liquid-phase continuous).

802.15 **efficiency** - Effectiveness of a component in transmitting or converting energy. For most mechanical drives, efficiency varies with both load and speed.

802.16 **feed point** - A nozzle or pipe connection through which material is fed either batchwise or continuously into the tank.

802.17 **flooding** - A condition where natural forces, typically buoyancy, dominate over imposed forces, such as mechanical mixing. Flooding is normally associated with gas dispersion, where too much gas may exceed the ability of the mixer to disperse the gas into the liquid.

802.18 **fluid force** - Lateral or axial force applied to the mixer impeller because of flow patterns or turbulence.

802.19 **harmonic ratio** - Ratio of operating speed, or any driving frequency to a natural frequency.
802.20 **immiscibility** - A condition in which two liquids fail to blend, and instead form two separate liquid phases with different properties and compositions. **Interfacial tension** exists along the surface separating the two phases. In typical liquid dispersions, one or the other phase is **continuous** and the other **dispersed** at given conditions.
802.21 **impeller** - The general term used to describe the device attached to the rotating shaft, which causes fluid motions. Impellers may be called by more specific terms, such as: propellers, turbines, etc.
802.22 **impeller power** - Actual power delivered to the process fluid by the impeller system.
802.23 **natural frequency** - First lateral vibrational frequency associated with an overhung shaft. Primary factors include impeller weight, shaft length, and diameter. Other components, including the supports, also have natural frequencies and other modes of vibration are possible and may be important in special cases.
802.24 **precision** - The ability to discriminate or resolve differences in measurements. Related to repeatability or the ability to reproduce a measurement within a defined tolerance.
802.25 **prime mover (motor)** - A device to convert primary energy to mechanical energy. Usually an electric motor in impeller mixer applications. Alternate prime movers include:

802.25.1 steam turbine
802.25.2 hydraulic motor
802.25.3 internal combustion engine
802.25.4 gas turbine

802.26 **pumping capacity** - A defined quantity used to characterize how much liquid motion is provided by the operation of a mixing impeller. Because typical impeller mixer operation involves only recirculating flows within a vessel, no readily available point-to-point measurement techniques apply.
802.27 **runout** - The variation in position caused by a rotating shaft that is not centered on the axis of rotation. The magnitude is usually measured as a peak-to-peak value.
802.28 **shaft** - The rotating extension on which impellers are mounted for mixing equipment. The shaft is typically of a cylindrical metal construction.

802.29 **shaft seal** - A device designed to retain pressure or restrict flow around a rotating shaft. Typical seal types include:

802.29.1 **lip seals** - simple elastomeric rings

802.29.2 **stuffing box seals** - compression packing materials around the shaft

802.29.3 **mechanical seals** - machined rings held against other machined surfaces, and a few other types.

802.30 **scale-up** - Techniques by which laboratory or pilot plant results are interpreted for use in large scale equipment. Similarity techniques are most often used. Besides geometric similarity, kinematic and dynamic similarity may be involved. The opposite process, called **scale-down,** may be done to conduct laboratory tests intended to duplicate large scale conditions.

802.31 **sparge ring** - Normally used for initial gas dispersion, air or whatever gas is brought into the tank through a pipe ring below the lowest impeller. The pipe has a series of holes drilled at equally spaced intervals to spread the gas to all blades of the impeller simultaneously.

802.32 **steady bearing** - A device that serves as a rotating bearing at the lower end or a mixer shaft. The steady bearing is normally attached to the bottom of the tank. A steady bearing increases the natural frequency and distributes hydraulic loads so a mixer can operate with a smaller diameter shaft than would be required for an overhung shaft.

802.33 **structural resonance** - A condition that can occur when the operating vibration frequency is too close to the system natural frequency.

802.34 **suspension** - The action or result of combining a particulate solid and a liquid, and providing sufficient motion to retain fluid characteristics.

802.35 **tank** - Most often, a vertical cylindrical container. Typically fabricated of metal, perhaps closed at the ends by shaped heads. Design for use with an impeller-type mixer is usually stronger than for simple storage purposes.

802.36 **top-entering** - The most common mounting configuration for impeller-type mixing equipment. Center mounted is most common for large mixers, angle mounted, for small mixers. Other mounting arrangements include:

802.36.1 **side-entering** - Often used on field-erected tanks, storage tanks, and chests.

802.36.2 **bottom-entering** - Occasionally used as an alternative to top-entering, usually because of severe limitations for alternative mountings.

802.37 **vessel** - Another name for a tank, but most often used when referring to one designed for a specified pressure and temperature, and certified under the ASME Code.

802.38 **vibration** - The cyclical or oscillatory motion of a piece of equipment. Depending on frequency, vibrations in mixing equipment may appear as noise, shaking or rocking motions. Causes are numerous and may suggest mechanical problems.

802.39 **withdrawal point** - A nozzle or pipe connection through which material is removed either batchwise or continuously into the tank.

803.0 *Notation*

Generalized dimensions are given for the notation because both English and SI Metric are used in the examples. Generalized dimensions include: force [F], length [L], mass [M], temperature [T], and time [t].

a = specific area, $[L^2/L^3]$
C = off-bottom location of lower impeller, [L]
C_B = baffle clearance to tank bottom, [L]
C_i = clearance between impeller and tank, [L]
C_p = heat capacity, [FL/(MT)]
C_W = baffle clearance to tank wall, [L]
D = impeller diameter, [L]
D_{AB} = molecular diffusivity, $[L^2/t]$
D_S = swept impeller diameter, [L]
D_d = disk diameter, [L]
d = shaft diameter, [L]
d_p = particle diameter, [L]
E_V = electrical voltage, [V]
E_Y = modulus of elasticity, $[F/L^2]$
E_m = mechanical efficiency
e = runout, [L]
F = force, [F]
F_c = centrifugal force, [F]
F_f = lateral fluid force on an impeller, [F]
F_t = fluid force plus centrifugal, [F]

F_{th} = axial-flow impeller thrust, [F]
g = acceleration of gravity, [L/t^2]
g_c = gravitational constant, [ML/(Ft^2)]

Force	Mass	g_c	
N	kg	1	(kg / N) (m / s^2)
lb$_f$	lb$_m$	32.17	(lb$_m$ / lb$_f$) (ft /s ^{2})
kg$_f$	kg	1	(kg / kg$_f$)(m / s^2)
dyne	gram	1	(gm / dyne) (cm / s^2)
lb$_f$	slug	1	(slug / lb$_f$) (ft / s^2)

H = liquid depth, [L]
H_i = anchor or helix impeller height, [L]
h = heat transfer coefficient, [F/(LtT)]
h_{cord} = cord height (hydrofoil blade), [L]
I = moment of inertia, [L^4]
I_A = electrical current, [A]
k = thermal conductivity, [F/(tT)]
k_L = mass transfer coefficient, [L/t]
$k_L a$ = specific mass transfer coef., [1/t]
L = shaft length (below mounting), [L]
L_B = baffle length, [L]
L_b = bearing spacing, [L]
L_f = motor loading fraction
L_i = shaft length to an impeller, [L]
L_t = total shaft length, [L]
M_B = bending moment, [FL]
N = rotational speed, [rev/t]
N_c = critical speed (nat. freq.), [rev/t]
N_R = revolutions
n = scale-up exponent
O = shaft offset from tank centerline, [L]
P = power, [FL/t]
P_f = electrical power factor
P_{In} = motor input power, [FL/t]
P_m = motor power, [FL/t]
P_p = impeller power (process), [FL/t]

P_w = impeller power in water, [FL/t]
p = propeller or helix pitch, [L]
Q = impeller pumping capacity, $[L^3/t]$
Q_g = gas volumetric flow rate, $[L^3/t]$
R = radius, [L]
r = radius to center tip of impeller, [L]
S = impeller spacing, [L]
$S.G.$ = specific gravity
T = tank diameter, [L]
t = time, [t]
t_b = impeller blade thickness, [L]
t_{blend} = blending time, [t]
t_d = disk thickness, [L]
TIR = total indicated runout, [L]
V = fluid volume, $[L^3]$
v_t = terminal settling velocity, [L/t]
v_{ts} = impeller tip speed, [L/t]
W = impeller blade width (actual), [L]
W_b = baffle width, [L]
W_i = impeller weight, [M]
W_d = impeller blade width (developed), [L]
W_e = impeller blade width (expanded), [L]
W_p = impeller blade width (projected), [L]
w_s = shaft weight per unit length, [M/L]
X = mass fraction
Z = liquid depth, [L]

803.1 *Greek Letters*

α = blade angle, [deg]
α_{cord} = cord angle, [deg]
α, β = propeller shaft angles, [deg]
δ = deflection of the shaft at the impeller, [L]
μ = fluid viscosity, [M/(Lt)]
ρ = fluid density, $[M/L^3]$
ρ_m = metal density, $[M/L^3]$
ρ_p = process fluid density, $[M/L^3]$
ρ_s = solid particle density, $[M/L^3]$
ρ_w = water density, $[M/L^3]$
σ = surface tension, $[F/L^2]$
τ = torque, [FL]

803.2 *Dimensionless Groups*

Ae	=	aeration number	=	$Q_g/(ND^3)$
Th	=	axial thrust number	=	$F_{th}g_c/(\rho N^2D^4)$
Fo	=	fluid force number	=	$F_f g_c/(\rho N^2D^4)$
Fr	=	Froude number	=	N^2D/g
Nu	=	Nusselt number	=	hD/k
Pe	=	Peclet number	=	$C_p\rho ND^2/k$
Po	=	power number	=	$Pg_c/(\rho N^3D^5)$
Pr	=	Prandtl number	=	$C_p\mu/k$
Pu	=	pumping number	=	$Q/(ND^3)$
Re	=	Reynolds number	=	$D^2N\rho/\mu$
Sc	=	Schmidt number	=	$\mu/(\rho D_{AB})$
Sh	=	Sherwood number	=	k_LD/D_{AB}
Th	=	thrust number	=	$F_{th}g_c/(\rho N^2D^4)$
We	=	Weber number	=	$\rho N^2D^3/(\sigma g_c)$
Θ	=	blend time number	=	$t_{blend}N$

804.0 *Equations and Related Groups*

804.1 *Impeller Flow*

Discharge flow produced by an impeller can be estimated from empirical data for pumping numbers, see next section for some typical values:

$$Q = Pu\,N\,D^3$$

Q = impeller pumping rate
Pu = impeller pumping number
N = rotational speed
D = impeller diameter

804.1.1 English units

$$Q = 4.33\text{x}10^{-3}\,Pu\,N\,D^3$$

Q [gal/min]
Pu [dimensionless]
N [rev/min]

D [inches]

804.1.2 SI metric units

$$Q = 0.159\, Pu\, N\, D^3$$

Q [m^3/s]
Pu [dimensionless]
N [rad/s]
D [m]

804.2 ***Power and Pumping Numbers***

Some typical values of *Po* and *Pu** for various impellers are provided for basic guidance. Because of design and application details, a range of values may exist for each category. Interpretation of pumping number varies, and pumping capacities depend on the impeller to tank size.

	Po	*Pu*
Axial flow:		
Propeller 1.0 : 1 pitch	0.3	0.4
1.5 : 1 pitch	0.8	0.6
45° pitched-blade turbine **		
4 blade ($W/D = 0.15$)	1.3	0.6
4 blade ($W/D = 0.20$)	1.7	0.8
Radial flow:		
Straight-blade turbine		
4 blade ($W/D = 0.15$)	3.2	1.1
4 blade ($W/D = 0.2$)	4.3	1.3
Disc turbine		
4 blade ($W/D = 0.2$)	5.4	1.4
6 blade ($W/D = 0.25$)	6.6	1.3

* Data assume turbulent flow with no significant wall interference effect. *Po* is based on impeller power. *Pu* is based on primary flow, rather than total induced flow.

** Width to diameter ratio is projected width divided by actual diameter.

Axial flow (hydrofoil) impellers are a major new class of impellers, however geometries vary significantly between suppliers. Power numbers between 0.1 and 1.0 and pumping numbers between 0.3 and 0.7 are possible.

805.0 *Calculation Examples*

805.1 *Power*

805.1.1 Wattmeter

A wattmeter is attached to a 100-HP electric motor-driven mixer.

P_{In} = 90 kW Start-Up
P_{In} = 73 kW Steady State

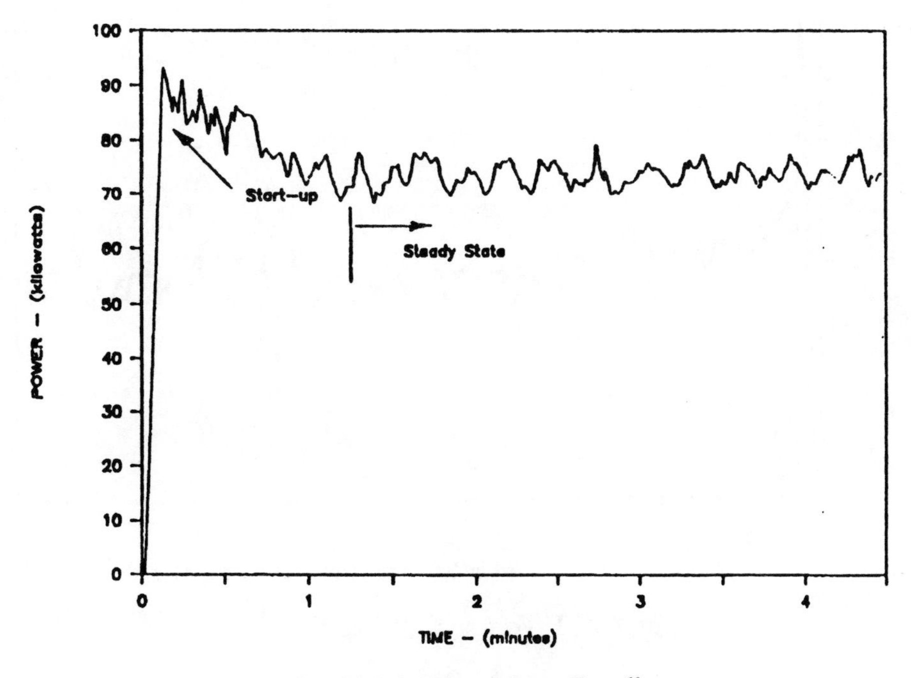

Fig. 805.1 Watt Meter Reading

805.1.2 **Amp Probe and Volt Meter**

Steady Operating Condition

I_A = 105 amps
E_V = 460 volts
P_f = 0.86
Three Phase Power

P_{In} = $105 \times 460 \times 0.86 \times \sqrt{3}$
= 71.9 kW

805.1.3 **Torque - Rotating Torque Cell**

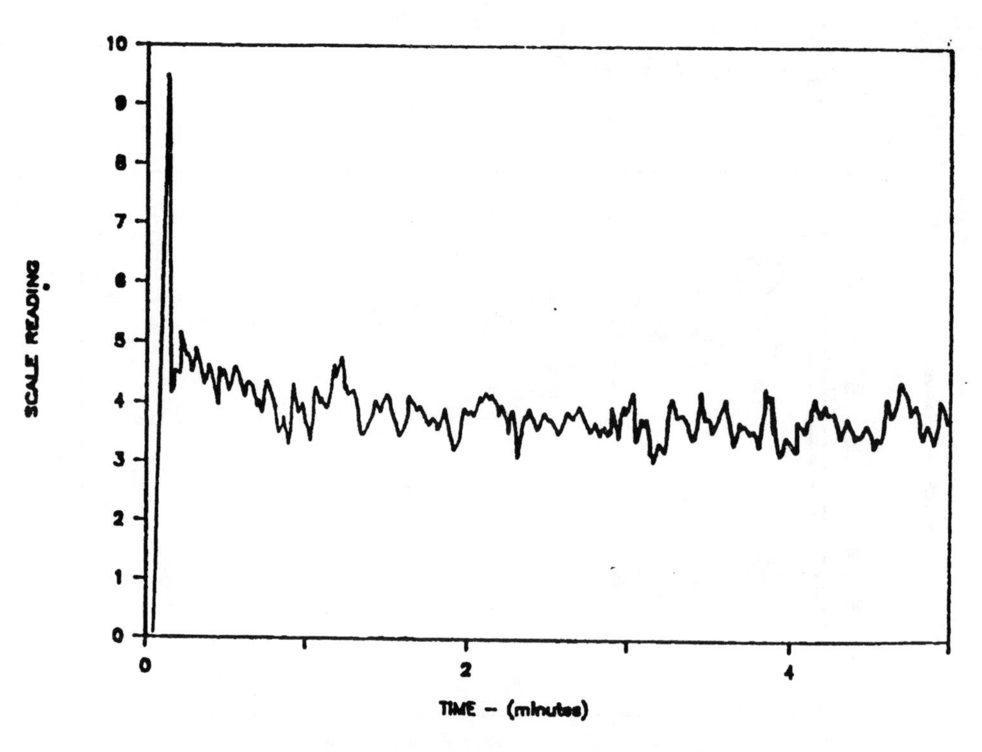

Fig. 805.2 Torque Cell Reading

A 10,000 inch-pound capacity torque cell is mounted between the gear drive input and electric motor output shaft. Shaft speed was determined using a stroboscope and a mark on the motor output shaft.

NOTE: Unit is calibrated by inserting resistor into bridge. Torque equivalent to a reading as given by the manufacturer: six units equivalent to five thousand inch-pounds.

Start-Up: Reading = 4.9
Speed = 1743 rpm

τ = (4.9/6) x 5000 = 4083 in-lb$_f$

P_m = $1.59 \times 10^{-5} \times 1743 \times 4083$

= 113 hp

Steady-State: Reading = 3.9
Speed = 1750 rpm

τ = (3.9/6)x5000 = 3250 in-lb$_f$

P_m = $1.59 \times 10^{-5} \times 1750 \times 3250$

= 90.4 hp

805.1.4 **Reaction Torque - Bearing Mounted**

A small laboratory mixer is mounted on a low friction bearing. Restraining the rotation of the mixer is a lever arm attached to a force gauge. A 6-inch diameter turbine impeller is run at several different speeds in water. The speeds and force readings are to be used to calculate power and power number.

Impeller Diameter = 6 inches
Lever Arm = 3.10 inches
Rotational Speed = N rpm
Scale Reading = F lb$_f$

Torque

$$\tau = 3.10 \times F \text{ inch-pounds}$$

Power (from Sec. 602.3.1)

$$P = \tau N / 63025$$
$$\tau = 3.10 \times F \times N / 63025$$
$$= 4.92 \times 10^{-5} N F \text{ (hp)}$$

Power Number (from Sec 602.2.1)

$$N_P = \frac{1.52 \times 10^{13}\ P}{S.G.\ N^3\ D^5}$$

$$= \frac{1.52 \times 10^{13}\ P}{1.0\ N^3\ 6^5}$$

$$= 1.95 \times 10^9\ P \,/\, N^3$$

N(rpm)	*F*(lb_f)	*P*(hp)	*Po*
310	5.3	0.081	5.3
495	14.3	0.335	5.6
690	28.0	0.95	5.6

805.1.5 **Steady State Power**

Calculate the steady state mixer power based on the watt meter reading in Fig. 805.1 and the equipment shown in Fig. 805.3. The mixer consists of an electric motor, represented by the motor curve in Fig. 805.4, and a gear drive, represented in Fig. 805.5.

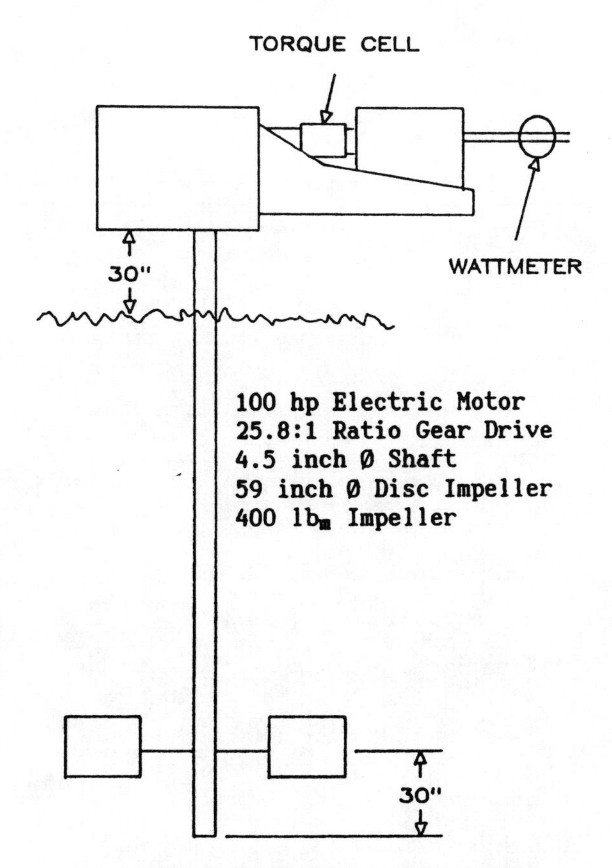

Fig. 805.3 Mixer for Sample Problems

P_{In}	=	73 kW	(Sec. 805.1.1)
P_m	=	90 hp	(Fig. 805.4)

Gear drive efficiency

$$E_m = \frac{90 - 6.0}{90} \quad \text{(Fig. 805.5)}$$

$$= 0.93$$

Mixer Power

$$P = 0.93 \times 90 \text{ hp}$$

$$= 83.5 \text{ hp}$$

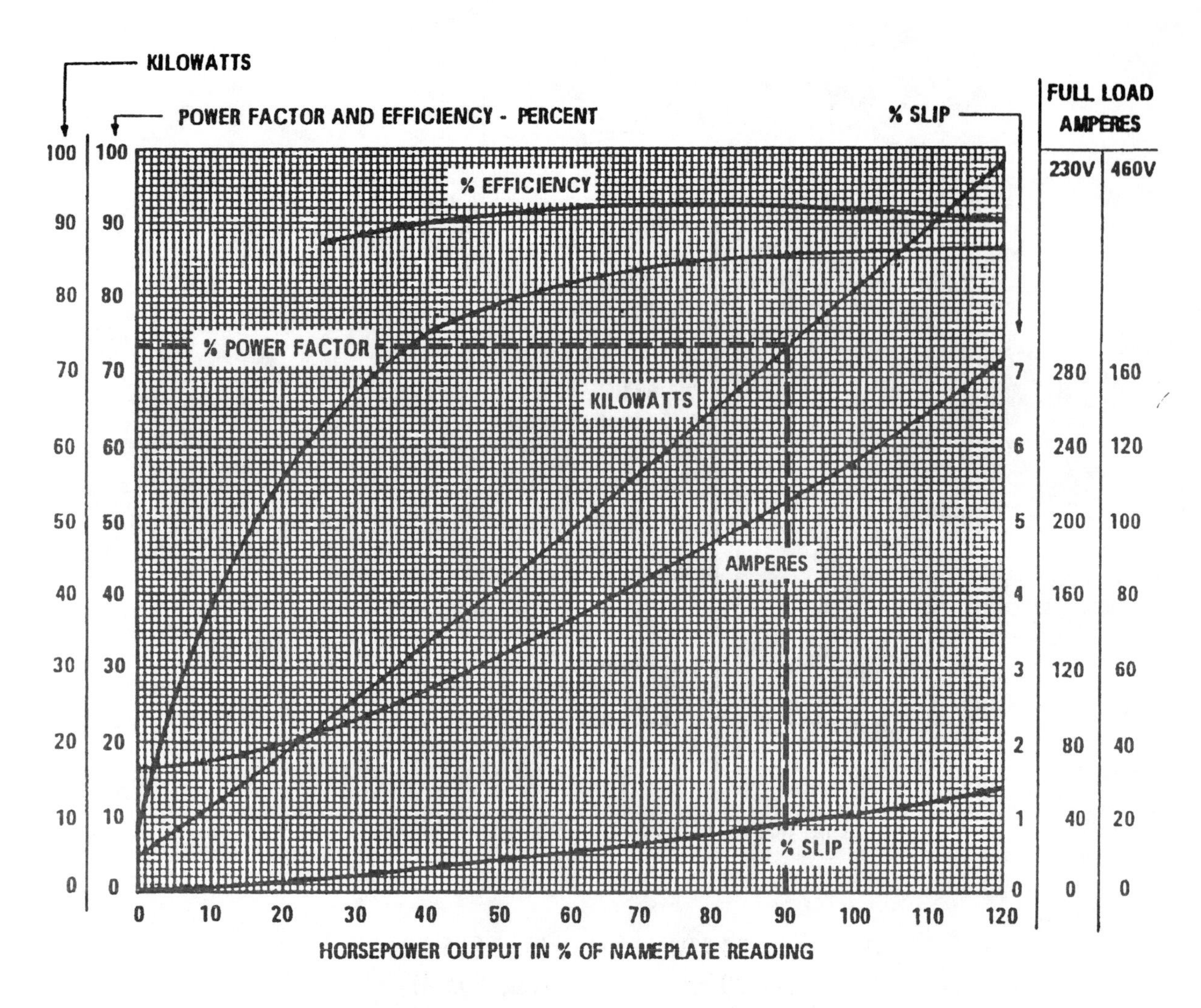

Fig. 805.4 Typical Motor Performance Curve
Horsepower 100, Frame 405T, R.P.M. 1780,
Cycle 60, Phase 3, Volts 230/460, Full Load Amps 234/117

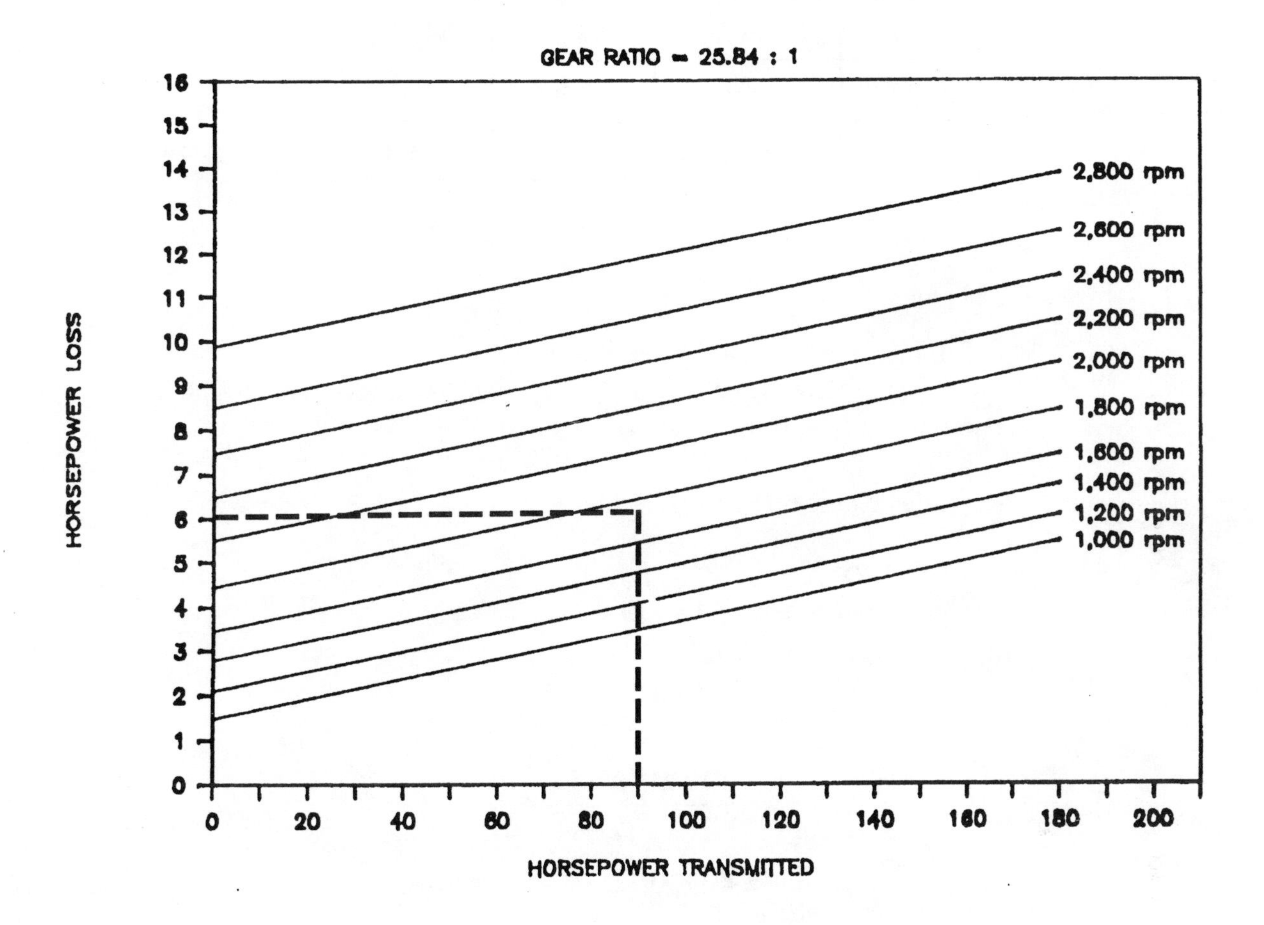

Fig. 805.5 Typical Power Loss for Mixer Drive

805.1.6 **Speed**

While a tachometer or stroboscope is necessary to measure motor shaft speed, the output shaft speed on many mixers is low enough to count and time several revolutions. The following data shows a simple determination of rotational speed:

For high temperature or high speed operation, a seal fluid recirculation system, with an external heat exchanger may be required to keep the seal at a safe operating temperature. A test of the seal system may be an appropriate part of the mixer test.

Rev	Time	rpm	(cps)
50	0:44.3	67.7	(1.13)
80	1:11.6	67.0	(1.12)
60	0:52.7	68.3	(1.14)

$$N = 67.6 \pm 0.70$$

805.1.7 **Theoretical Mixer Power**

The mixing impeller is a 60-inch diameter, disk-style turbine operating at 67.6 rpm.

Reynolds Number (Sec. 602.1.1):

$$D = 59 \text{ inches}$$
$$N = 67.6 \text{ rpm}$$
$$S.G. = 1.0$$
$$\mu = 1.0 \text{ cp}$$

$$N_{Re} = \frac{10.7\, D^2\, N\, (S.G.)}{\mu}$$
$$= 2.5 \times 10^6$$

Thus, the impeller is operating in the turbulent region. Reported power numbers range from 5.0 to 5.7. Rearranging the power number from Sec. 602.2.1:

$$P\,[\text{hp}] = 6.55 \times 10^{-14}\, Po\, (S.G.)\, N^3\, D^5$$

Thus we would expect:

$$78.7 < P\,[\text{hp}] < 89.7$$

In Sec. 805.1.5 we calculated 83.5 hp to the mixer based on measured power and correcting for component efficiencies. The two values for impeller power show excellent consistency.

For more information on power and power numbers, most comprehensive references on mixing provide both standard and special values.

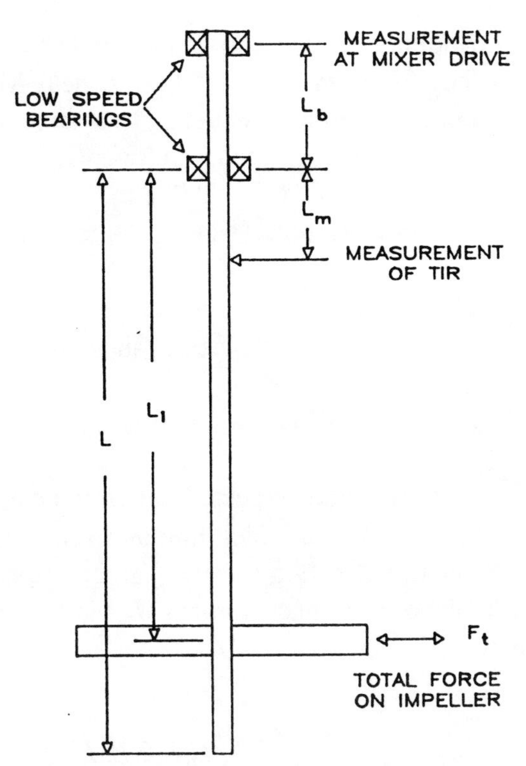

Fig. 805.6 Mixer Shaft for Sample Problems

805.2 *Mechanical*

805.2.1 Dynamic Shaft Deflection

A mixer vibration problem has been observed during commissioning. The important shaft dimensions are shown in Fig. 805.6.

Bearing Spacing,	L_b	=	24 inches
Overhung Length,	L	=	240 inches
Length to Impeller,	L_i	=	210 inches
Shaft Diameter,	d	=	4.5 inches
Impeller Weight,	W_i	=	400 lbs.

The vessel was filled with water for these tests. Dial gauge measurements were made for motion or total indicated runout (TIR) as follows:

1. Support structure - no measurable flexure.
2. Top of mixer - TIR = 0.010 inches. Mounting fasteners checked and retightened. No change in motion.
3. Mixer shaft: 15 inches below drive
   Static runout: TIR = 0.008 in.
   Dynamic runout: TIR = 0.160 in.

A rough estimate of deflection due to shaft dynamics is half the difference between dynamic and static runout:

(0.160 - 0.008) / 2 = 0.076 inches

805.2.2 Calculate the expected deflection caused by a 350 pound fluid force (F_f) and 1 inch runout. From Fig. 805.6, the equation for shaft deflection versus F_t at the impeller location is obtained. The point of measurement (L_m) is 15 inches below the lower bearing.

$$\delta = \frac{F_t\, L_m}{6\, E_Y\, I}\left(2\, L_i\, L_b + 3\, L_i\, L_m + L_m{}^2\right)$$

Modulus of Elasticity

$$E_y = 28.5 \times 10^6 \quad \mathrm{lb_f/inch^2}$$

Moment of Inertia

$$I = \frac{\pi}{64} \times 4.5^4 = 20.13 \quad \mathrm{inch}^4$$

Calculating for Lengths

$$\delta = 8.41 \times 10^{-5}\, F_t$$

Forces

Centrifugal force

$$F_c = \left(2\pi\, N/60\right) 2W\delta/g_c$$

Runout,	$\delta = 1$ inch
Impeller Weight,	$W = 400\ lb_m$
Speed,	$N = 67.6$ rpm

$$F_c = 52\,lb_f$$

Centrifugal force due to runout at the impeller is a small factor compared with fluid forces.

The total force is at the impeller.

$$\begin{aligned} F_t &= F_f + F_c \\ &= 350 + 52 \\ &= 402\ lb_f \end{aligned}$$

Thus:

$$\begin{aligned} \delta &= 8.41 \times 10^{-5}\, F_t \\ &= 8.41 \times 10^{-5} \times 402 \\ &= 0.034 \text{ inches} \end{aligned}$$

Total indicated runout (TIR) is the sum of the runout at the measurement point plus twice the deflection. This multiplier of two accounts for deflections toward and away from the gauge.

$$\begin{aligned} \text{TIR} &= 0.008 + 2 \times 0.034 \\ &= 0.076 \text{ inches} \end{aligned}$$

The measured deflection (Sec. 805.2.1) is twice this value. Fluid forces are either much larger than predicted or a dynamic interaction between the shaft operating speed and the natural frequency exists.

805.2.3 A detailed physical inspection of the mixer of in this example revealed only one discrepancy with the drawing,

Fig. 805.6. The impeller was located at the bottom of the shaft. The drawing called for a placement 30 inches from the bottom of the shaft. Runout was subsequently measured to be 1 inch at the impeller.

805.2.4 Calculate natural frequency for the impeller located at the bottom of the shaft and 30 inches from the bottom. For a constant diameter shaft, Fig. 805.6.

Natural Frequency

$$N_C = 37.8 \frac{d_2 \sqrt{E_Y / \rho_m}}{L \sqrt{W_e} \sqrt{L + L_b}}$$

where:

Shaft Diameter,	d	=	4.5 inches
Modulus of Elasticity,	E_Y	=	28.5×10^6 psi
Metal Density,	ρ_m	=	0.284 lb_m/in^3
Shaft Length,	L	=	240 inches
Bearing Spacing,	L_b	=	24 inches

Equivalent Weight at End of Shaft (L)

$$W_e = W_i \left(L_i / L\right)^3 + w_s \; L/4$$

Impeller Weight,	W_i	=	400 lb_m
Shaft Unit Weight	w_s	=	4.45 lb_m/in

Calculation for Impeller at Bottom of Shaft:

$$\begin{aligned} W_e &= 400\left(240/240\right)^3 + 4.45\left(240/4\right) \\ &= 400 + 267 \\ &= 667 \end{aligned}$$

$$N_C = 37.8\frac{4.5^2\sqrt{28.5\times10^6/0.284}}{240\sqrt{667}\sqrt{240+24}}$$
$$= 76.1\,\text{rpm}$$

Operating Speed, $N = 67.6$ rpm

$$N/N_C = 67.6/76.1$$
$$= 0.89$$

With the shaft operating at 89 percent of critical, serious problems exist, and the high deflections are probably caused by this condition.

Calculation for Impeller 30 inches from the Bottom of Shaft:

$$W_e = 400\left(210/240\right)^3 + 4.45\left(240/4\right)$$
$$= 400\left(0.67\right) + 267$$
$$= 535\,\text{lbs}$$

$$N_C = 37.8\frac{4.5^2\sqrt{28.5\times10^6/0.284}}{240\sqrt{535}\sqrt{240+24}}$$
$$= 85.0\,\text{rpm}$$

Operating Speed, $N = 67.6$ rpm

$$N/N_C = 67.6/85.0$$
$$= 0.80$$

With the shaft operating at 80 percent of critical, a reasonable design condition exists.

The mixer was retested with the impeller properly located 30 inches from the bottom of the shaft, deflections measured 15 inches below the drive were as follows:

Static runout,	TIR	=	0.008 inches
Dynamic runout,	TIR	=	0.050 inches

Problems solved.

805.3 ***Liquid-Solid***

805.3.1 Nearly spherical particles of 120 microns (micrometer) diameter are to be suspended in a fluid of 0.008 Pa·s (8 cp) viscosity and 0.95 specific gravity. The absolute particle specific gravity is 3.2 (density of 3200 kg/m^3). Estimate the terminal settling velocity of the particles from Stokes Law.

From Stokes Law, the terminal settling velocity (v_t) is calculated:

$$v_t = \frac{(\rho_s - \rho)\, d_p^{\,2}\, g}{18\,\mu}$$

$$= \frac{(3200 - 950\ \mathrm{kg/m^3}) \times 0.00012\ \mathrm{m} \times 9.806\ \mathrm{m/s^2}}{18 \times 0.008\ \mathrm{Pa{\bullet}s}}$$

$$= 0.0022\ \mathrm{m/s} = 0.43\ \mathrm{ft/min}$$

If the particle density is low enough that the particles do not interact extensively, settling rates can be determined based on individual particle settling velocities. The Stokes Law correlation is accurate at particle Reynolds numbers below 0.1 and reasonably accurate between 0.1 and 1, which is based on the fluid viscosity and density and the particle diameter. For this example, the particle Reynolds number ($Re_{Particle}$) is

$$\mathrm{Re}_{Particle} = \frac{d_p\, \rho\, v_t}{\mu}$$

$$= \frac{(0.00012\ \mathrm{m})(950\ \mathrm{kg/m^3})(0.0022\ \mathrm{m/s})}{0.001\ \mathrm{Pa{\bullet}\,s}}$$

$$= 0.031$$

Since this is less than 0.1, the Stokes Law correlation is valid.

806.0 *Sample Log Sheet*

The purpose of this log sheet is to provide a guide for recording of data required to conduct and analyze a performance test for specific mixing situations.

806.1 *System Test Log*

A complete record of all conditions and operations performed during testing is essential. The results of some tests may have to be analyzed later.

806.2 *Physical Description Sketch*

A complete sketch of the system, including variations in liquid levels, will frequently help identify potential problems.

806.3 *General Process Log*

The following page provides a general data sheet for recording test results. The sheet should be modified for specific applications.

MIXING TEST -- GENERAL PROCESS LOG SHEET

Test Name ______________________________ Ident. No. ______________

Location ______________________________ Date ______________

Test Operator ______________________________

AMBIENT CONDITIONS Temp. __________ Bar. Pres. __________ Humidity __________

FLUIDS--MATERIALS

Material	Quantity	Sp. Gr.	Viscosity[1]	Other
Mixture				

[1] If non-Newtonian, identify type and how viscosity was measured under Notes.

Solids--attach size distribution, sieve analysis, general description, lot number, etc.

VESSEL (Sketch total system to scale and show all relevant dimensions)

Diameter __________ Vessel height __________ Top/bottom heads __________

Baffles number/size/location ______________________________

Inlets/outlets/coils/obstructions ______________________________

MIXER

Type/Model __________ Power __________ Output speed __________

Impellers number/type/size ______________________________

Impeller location ______________________________

OPERATING CONDITIONS

Temperature __________ Pressure __________ Flow __________

Liquid Levels: Min. __________ Normal __________ Max. __________

Liquid Volumes: Min. __________ Normal __________ Max __________

Describe mixing cycle on back.

TEST LOG

Time	Sample No.	Flow			Temp.	Mixer Speed	Power Draw	Measured Results	Comments
		1	2	3					

807.0 *Scale-Up/Scale-Down*

The purpose of many mixing tests is to give assistance with scale-up or scale-down of a process. Scale-up is the more common concern to produce production scale equipment. Scale-down, however, can be as large a concern. To do scale-down, the mixing tests must be designed before scale-up, so the pilot scale equipment simulates production scale performance.

Different classes of mixing problems are controlled by different hydrodynamic regimes, and different types of correlations have been developed. No universally accepted procedures or correlations exist for handling scale-up tests. Below are some guidelines that have frequently been used. In addition it is highly recommended to contact personnel experienced in mixing processes and mixing scale-up for assistance.

807.1 *Geometric Similarity*

During scale tests, maintaining geometric similarity for all mechanical components between mixing systems is often useful, including:

807.1.1 Impeller style and number

807.1.2 Number and style of baffles

807.1.3 Geometry ratios such as the impeller-to-tank-diameter, fluid-height-to-diameter, etc.

807.2 *Scale Parameter*

A scale parameter should be selected, such as:

807.2.1 A linear dimension, such as the vessel diameter for a circular cross-section vessel.

807.2.2 Vessel volume

807.3 *Correlating Process Parameters*

A process parameter must be selected to use in scale-up. Measurements of it can be made and correlations produced. These parameters will be appropriately modified and not necessarily kept constant on scale-up. With proper mathematical manipulations, correlations can frequently be changed from one process parameter to another.

It should be noted that in scale-up or scale-down, keeping all parameters constant is impossible, so one must determine which parameters are most important. Assistance from an experienced mixing expert is helpful in selecting which parameter to use.

Frequently used correlating parameters include:

807.3.1 Power per volume - P/V

807.3.2 Torque per volume - τ/V

807.3.3 Impeller tip speed - ND
807.3.4 Bulk fluid velocity or velocity at a point
807.3.5 Arbitrary combinations of basic variables, such as:

$$N^{x_1} \times D^{x_2} \times T^{x_3}$$

807.4 ***Correlating Conditions***

When a scale test is being run, one must define what conditions are to be compared (remain constant) to determine when equal performance is reached. Many such conditions exist, depending on the process needs. Examples of potential correlating conditions are:

807.4.1 A visual or measured degree of uniformity.
807.4.2 A visual or measured degree of solids suspension.
807.4.3 A gas dispersion rate or superficial gas velocity.
807.4.4 A desired blend time.
807.4.5 A fluid velocity measured at a significant point in the vessel.
807.4.6 The type of surface motion.
807.4.7 A reaction rate or rate coefficient.
807.4.8 A standard fluid against which other produced fluids may be compared in quality.

807.5 ***Mixing Correlation***

Usually the result of the scale test is a scale equation. The testing requires evaluating the form of the correlation and determining the value of appropriate scale factors (x_1, x_2, x_3). Scale equations can take various forms including:

$$\left(\frac{P}{V}\right)_2 = \left(\frac{P}{V}\right)_1 \left(\frac{T_2}{T_1}\right)^{x_1}$$

$$N_2^{x_1} D_2^{x_2} T_2^{x_3} = N_1^{x_1} D_1^{x_2} T_1^{x_3}$$

$$\left(\frac{\tau}{V}\right)_2 = \left(\frac{\tau}{V}\right)_1 \left(\frac{V_2}{V_1}\right)^{x_1}$$

$$N_2 = N_1 \left(\frac{D_1}{D_2} \right)^n$$

808.0 *References*

808.1 ASME Power Test Code, Supplement, Instruments and Apparatus. Part 2. Pressure Measurement, PTC 19.2-1964.
808.2 Ibid., Part 3. Temperature Measurement, PTC 19.3-1974.
808.3 Ibid., Part 5. Part II of Fluid Meters, PTC 19.5-1971.
808.4 Ibid. Part 7. Measurement of Shaft Power, PTC 19.7-1980.
808.5 Ibid. Part 16. Density Determinations of Solids and Liquids, PTC 19.16-1965.
808.6 Ibid. Part 17. Determination of the Viscosity of Liquids, PTC 19.17-1965.
808.7 Uhl, V. W., and J. B. Gray, Editors, **Mixing, Theory and Practice,**
Vol. I, Academic Press, 1966
808.7.1 Introduction
808.7.2 Fluid Motion and Mixing
808.7.3 Impeller Characteristics and Power
808.7.4 Flow Patterns, Fluid Velocities, and Mixing in Agitated Vessels
808.7.5 Mechanically Aided Heat Transfer
Vol. II, Academic Press, 1967
808.7.6 Mass Transfer
808.7.7 Mixing and Chemical Reactions
808.7.8 Mixing of High Viscosity Materials
808.7.9 Suspension of Solids
808.7.10 Mixing of Solids
808.7.11 Mechanical Design of Impeller-Type Liquid Mixing Equipment
Vol. III, Academic Press, 1986
808.7.12 Agitation of Particulate Solid-Liquid Mixtures
808.7.13 Turbulent Radial Mixing in Pipes
808.7.14 Flow and Turbulence in Vessels with Axial Impellers
808.7.15 Scale-Up of Equipment for Agitating Liquids
808.7.16 Mixing of Particulate Solids

808.8 Comstock, J. E. (editor), **Principles of Naval Architecture**, The Society of Naval Architects and Marine Engineers (SNAME), 1967.

808.9 Nagata, S., **Mixing, Principles and Applications**, Halsted Press, John Wiley & Sons, 1975.

808.10 Gates, L. E., D. S. Dickey, R. W. Hicks, et. al, **CE Refresher - Liquid Agitation, Chemical Engineering**:

- 808.10.1 How to select the optimum turbine agitator, Dec. 8, 1975, pp 110-114.
- 808.10.2 Dimensional analysis for fluid agitation systems, Jan. 5, 1976, pp 139-145.
- 808.10.3 Fundamentals of agitation, Feb. 2, 1976, pp 93-100.
- 808.10.4 How to design agitators for desired process response, Apr. 26, 1976, pp 102-110.
- 808.10.5 Selecting agitator systems to suspend solids in liquids, May 24, 1976, pp 144-150.
- 808.10.6 How to select agitators for dispersing gas into liquids, July 19, 1976, pp 141-148.
- 808.10.7 How to specify drive trains for turbine agitators, Aug. 2, 1976, pp 89-94.
- 808.10.8 How the design of shafts, seals and impellers affects agitator performance, Aug. 30, 1976, pp 101-108.
- 808.10.9 Cost estimation for turbine agitators, Sept. 27, 1976, pp 109-112.
- 808.10.10 How to use scale-up methods for turbine agitators, Oct. 25, 1976, pp 119-126.
- 808.10.11 Applications analysis for turbine agitators, Nov. 8, 1976, pp 127-133.
- 808.10.12 Application guidelines for turbine agitators, Dec. 6, 1976, pp 165-170.

808.11 AIChE Equipment Testing Procedure, **Dry Solids, Paste and Dough Mixing Equipment**, J. B. Gray, Chairman, AIChE, 1979.

808.12 Oldshue, J. Y., **Fluid Mixing Technology**, Chemical Engineering, McGraw-Hill Publication Co., 1983.

808.13 Harnby, N., M. F. Edwards, and A. W. Nienow, **Mixing in the Process Industries**, Butterworths & Co., Ltd, 1985.

808.14 Tatterson, G. B., **Fluid Mixing and Gas Dispersion in Agitated Tanks**, McGraw-Hill, Inc., 1991.

808.15 Tatterson, G. B., **Scaleup and Design in Industrial Mixing Processes**, McGraw-Hill, Inc., 1994.

808.16 Chopey, N. P., **Handbook of Chemical Engineering Calculations**, Second Edition, Section 12. Liquid Agitation, McGraw-Hill Book Co., 1994.

808.17 Fasano, J. B., Bakker, A., Penney, W. R., Advanced Impeller Geometry Boosts Liquid Agitation, **Chemical Engineering**, August 1994, pp. 110-116.

808.18 Corpstein, R. R., Fasano, J. B., Myers, K. J., The High-Efficiency Road to Liquid-Solid Agitation, **Chemical Engineering** , October 1994, pp. 138-144.

808.19 Bakker, A., J. M. Smith, and K. J. Myers, How to Disperse Gases in Liquids, **Chemical Engineering**, December 1994, pages 98-104.

808.20 Bakker, A., Morton, J. R., Berg, G. M., Computerizing the Steps of Mixer Selection, **Chemical Engineering**, March 1994, pp. 120-129.

808.21 Bakker, A., Fasano, J. B., Leng, D. E., Pinpoint Mixing Problems with Lasers and Simulation Software, **Chemical Engineering**, January 1994, pp. 94-100.

808.22 Fasano, J. B., Miller, J. L., Pasley, S. A., Consider Mechanical Design of Agitators, **Chemical Engineering Progress**, August 1995, pp. 60-71.

808.23 Liptak, B., Editor-in-Chief, **Instrument Engineers' Handbook**, Third Edition, Chilton Book Co., 1995.

808.24 Perry, R. H., D. W. Green, and J. O. Maloney, Editors, **Perry's Chemical Engineers' Handbook**, Seventh Edition, Chapter 18, Liquid-Solid Operations and Equipment,, Equipment, McGraw-Hill Book Co., 1997.

808.25 Dream, R. F., Heat Transfer in Agitated Jacketed Vessels, **Chemical Engineering**, January 1999, pp. 90-96.

Index